Biometrical Interpretation

D1323059

Biometrical Interpretation

Making Sense of Statistics in Biology

Second Edition

NEIL GILBERT

Oxford Melbourne New York Tokyo
OXFORD UNIVERSITY PRESS
1989

Oxford University Press, Walton Street, Oxford OX2 6DP
Oxford New York Toronto
Delhi Bombay Calcutta Madras Karachi
Petaling Jaya Singapore Hong Kong Tokyo
Nairobi Dar es Salaam Cape Town
Melbourne Auckland
and associated companies in
Berlin Ibadan

Oxford is a trade mark of Oxford University Press

Published in the United States
by Oxford University Press, New York

First published 1973
Second edition 1989

British Library Cataloguing in Publication Data
Gilbert, Neil
Biometrical interpretation. — 2nd ed.
1. Biology. Applications of statistical mathematics
I. Title
574'.01'5195
ISBN 0-19-854250-X

Library of Congress Cataloging in Publication Data
Gilbert, Neil
Biometrical interpretation.
Bibliography
Includes index.
1. Biometry. I. Title,
QH323.5.G54 1989   574'.028   88-25458
ISBN 0-19-854250-X (pbk.)

Typeset by Cotswold Typesetting Ltd, Cheltenham
Printed in Great Britain
at the University Printing House, Oxford
by David Stanford,
Printer to the University

Preface

This book is addressed to biologists who use biometrical methods as a tool. It is not the usual book of statistical recipes. These days, we have computers to do the donkey work; the biologist need not know how to calculate, say, a regression coefficient. But he does need, more than ever, to understand the use and interpretation of regressions. This book concentrates, not on how to do an analysis, but on how to choose the right sort of analysis, and on how to make sense of the answer. It deals with questions that often arise in practice. So it is written in self-defence, to answer questions which biologists ask again and again.

The first three chapters deal with basic statistical methods. Some parts are unavoidably complex. If you can cope with those three chapters (and with their examples) you have nothing to fear from statistics. The going is easier in later chapters. Chapters 10 and 11 are about two fields of biology which rely heavily on mathematics and statistics.

R. A. Fisher, the founder of modern statistics, once defined variance as 'the attitude of one statistician to another'. We can look at any particular question from several points of view. Statistical methods, apparently quite unrelated to each other, are in fact different aspects of the same central theory. Not everyone will agree with my point of view, which is roughly that of the Rothamsted school. The two conflicting quotations, both by physicists, underline the basic philosophy: that statistical methods constitute a tool, often useful but sometimes abused.

There has been no basic change in statistical methods since the first edition of the book was published in 1973. Some new methods, made possible by computers but founded on existing principles, have been introduced. Biologists now use computers to proliferate statistical analyses, which have become an integral part of their work. But the intimate contact with the data—inescapable before the days of computers—has been lost, and sometimes an analysis is applied to data which contradict the assumptions of that analysis! The book has been revised accordingly.

N. E. G.

Hobart, 1988

Contents

1 Least squares

1.1. Remainders

Statistical methods are used to *describe* sets of data, and to *predict* (as well as possible) the values of other measurements which have not, or cannot, be made. The emphasis is on prediction. Nearly always, a set of data is only a sample; we are interested, not just in that sample, but in the whole population from which the sample is drawn. For example, we may have measurements on twenty Dalmatian dogs, but we wish to draw conclusions about Dalmatians—or perhaps even dogs—in general. We cannot expect those predictions to be perfectly accurate, but we want them to be as good as possible.

Suppose that we are interested in some measurement y. The value of y, measured on the ith dog, is y_i. We are going to divide the value y_i into two parts, F_i and a remainder. For F_i we choose some function whose value we can supply, and which will help us predict future values of y_i. Very often, F_i is a mean for all Dalmatians. Or F_i might represent a mean for male dogs and a mean for bitches, in which case we cannot supply its value in any given case until we know the sex of the animal concerned. Or F_i might be a regression of y_i on some other measurement x_i, so that we need to know x_i before we can supply F_i for any given animal. The basic principle is the same:

$$y_i = F_i + \text{remainder},$$

where F_i is a value which we can supply in any particular case, and which helps us to predict y_i; while the 'remainder', or 'residual', means 'a bit left over which we can't, or can't be bothered to, specify'. (The important assumption that the two parts add together will be re-examined in Chapter 4.) The same distinction between what we know (F_i) and what we don't know (remainder) occurs in the closely related subjects of statistical mechanics and information theory.

The remainder arises in two ways. It may be an error of measurement—we can never measure anything exactly, although we can *count* exactly. Or the remainder may represent something biological, which we can't—or can't be bothered to—predict. For example, the measurement y might be the number of eggs in a clutch of duck eggs. We may be content to say that a duck lays on average 4.6 eggs. Then $F_i = 4.6$ in every case: but no duck can lay exactly 4.6 eggs. If a duck actually lays five eggs, the residual is $+0.4$. We know that there are physiological processes which decide that a particular duck shall lay 1, 2, 3, 4 ... eggs, but we cannot specify—or perhaps are not interested in—the precise number for any given duck. The word 'error' is often used instead of 'remainder'. But the word error implies some kind of mistake. It suggests that the duck ought to lay 4.6 eggs. The terminology has sometimes trapped biologists into regarding genuine biological variation as a regrettable nuisance. I shall reserve the term 'error' for errors of measurement. The words 'remainder' and 'residual' mean the same thing, and will be used interchangeably.

1.2. Sum of squares

The value of F_i may sometimes be supplied theoretically, but usually it is estimated from a sample of data. In that case, we want the estimate to give the best possible predictions, i.e. to make the residuals as small as possible. It is up to us to specify what *kind* (mean, regression etc.) of function F_i should be: too bad if we choose an F_i which is incapable of predicting y_i. For example, we might suppose that some average m can predict future values of y. (I shall make the usual distinction between the unknown population mean m and its known estimate $\bar{y}$.) Then we shall have

$$y_i = m + \text{remainder}$$

and we want the estimate of m to make the remainders, in the whole population, as small as possible. We therefore use an estimate of m which makes the remainders in the observed sample small. Very often, we estimate m by minimizing $\sum(y_i - m)^2$—that is, the sum of squares of all the remainders in the sample. We might instead estimate m by minimizing some other combination of the sample remainders $(y_i - m)$. But the sum of squares offers many advantages. It leads to simple methods of calculation. It puts positive and negative remainders on an equal footing, and pays greater attention to large remainders than to small: for example, a remainder of -1 contributes 1 to the sum of squares, while a remainder of $+2$ contributes 4. There are also theoretical reasons (Chapter 7) for preferring the sum of squares. It can be shown that the estimate of m which minimizes $\sum(y_i - m)^2$ is the arithmetic mean $\bar{y}$ of the sample values of y. We use the fact that $\bar{y}$ minimizes the sum of squares of the remainders to *justify* our use of the arithmetic mean $\bar{y}$, rather than any other combination of the values of y. Similarly, if we suppose that y can be predicted by a linear regression on some known measurement x, so that $F_i = a + bx_i$, the customary estimates of a and b are those that minimize $\sum(y_i - a - bx_i)^2$, summed over all values of y_i (with concomitant x_i) in the sample.[1]

If the sample values of y are 31, 32, 29, 31, 30, 28, 30 and 29, it is reasonable to suppose that future values of y will also be somewhere near the sample mean 30.0, so that the equation '$y = 30.0 + \text{residual}$' is likely to predict values of y better than the equation '$y = 0 + \text{residual}$'. In other words, the sample mean is significantly different from zero. But if the values of y are 1, 2, -1, 1 and 0, it is doubtful if the sample mean gives a worthwhile prediction of other values of y. The mean is not significantly different from zero, and it does not pay—on the evidence of these data alone—to think in terms of a non-zero mean m. You may say 'Cannot a mean of zero predict y just as well as a mean of 30?' It can't. A mean of zero can *describe* the whole set of ys (Chapter 2 will examine the descriptive use of means) but to say 'I expect that $y = 0 + \text{an}$ unpredictable remainder' is just the same as saying 'I expect that $y = \text{something unpredictable}$'. So long as the emphasis is on prediction, a zero mean does not help.[2]

1.3. Significance tests

We not only use the sample values of y to estimate m, but we also use them to assess whether that estimate will indeed help us to predict other values of y. The significance test tries to determine whether the remainder $(y_i - \bar{y})$ will usually be smaller than y_i itself. If so, $\bar{y}$ gives useful prediction of y_i. Since $\bar{y}$ is estimated by minimizing $\sum(y_i - m)^2$, the question is whether $\sum(y_i - \bar{y})^2$ is much less than the original sum of squares $\sum y_i^2$. Unless $\bar{y}$ chances to be exactly zero, $\sum(y_i - \bar{y})^2$ will necessarily come out less than $\sum y_i^2$ since $\bar{y}$ is estimated by minimizing the sum of squares. So we ask 'How much smaller must $\sum(y_i - \bar{y})^2$ be, for $\bar{y}$ to be significant?'

To answer the question, we do not consider the sum of squares itself. It is here that the ideas of 'degrees of freedom' and 'mean square' arise. Whereas m is estimated by minimizing the sum of squares, we use the *mean* square to assess whether the estimation has been worthwhile. The expression 'mean square of y' means just that—it is a mean value of y^2. Suppose we draw a sample of N values of y from a population with mean zero and variance V. The variance is defined as the average value, in the population as a whole, of $(y_i - m)^2$: so in this case, when $m = 0$, the average value of y_i^2 is V. Then it can be shown (Example 2.8) that the value of y_i^2 in the sample will on average be NV, and that of $\sum(y_i - \bar{y})^2$ will on average be $(N-1)V$. If we say that $\sum y_i^2$ has N degrees of freedom and $\sum(y_i - \bar{y})^2$ has $(N-1)$ degrees of freedom, then the mean squares

$\sum y_i^2/N$ and $\sum(y_i - \bar{y})^2/(N-1)$ are both expected to equal the variance, provided that the population mean m is zero, i.e.

$$\frac{\sum y_i^2}{N} \quad \text{and} \quad \frac{\sum(y_i - \bar{y})^2}{N-1} \quad \text{on average equal } V.$$

$$\underset{\substack{\text{original} \\ \text{mean square}}}{} \qquad \underset{\substack{\text{remainder} \\ \text{mean square}}}{}$$

But if the remainder mean square is smaller than the original mean square, *either* it is smaller by chance *or* $\bar{y}$ (the estimate of m) can usefully predict values of y.[3]

For technical reasons, the variance-ratio (or F-) test does not directly compare the original and remainder mean squares. Instead, it compares the size of the *reduction* in variance achieved by fitting $\bar{y}$, with the size of the remainder mean square. (By 'fitting $\bar{y}$' we mean simply the process of calculating $\bar{y}$ and then subtracting it from each y_i to calculate the residuals.) The two comparisons are equivalent. The analysis of variance splits the original sum of squares into two parts: (1) the remainder sum of squares and (2) the difference, accounted for by the fitting of $\bar{y}$, between the original and remainder sums of squares. These two parts of the original total sum of squares correspond exactly to the two parts (remainder and F_i) of y_i. Both parts of the sum of squares are then converted to mean squares by dividing each by its appropriate degrees of freedom—in this case the remainder sum of squares by $(N-1)$ and the 'sum of squares due to fitting the mean' by 1.

If the population mean is zero, these two mean squares are expected to be equal, since they both estimate a purely residual variance. (You may object that if m is zero, it can't account for any of the variance, i.e. the sum of squares due to fitting the mean ought to be zero. But we can never know m exactly: we have to use the sample mean $\bar{y}$ instead, and $\bar{y}$ will not in general be zero, even when the population mean $m = 0$. As mentioned above, the fitting of $\bar{y}$ always produces *some* reduction in the 'residual' sum of squares, except in the rare case where $\bar{y}$ chances to be precisely zero.) The variance-ratio test then compares the 'mean square due to fitting the mean' with the remainder mean square. But the essential question is, does $\bar{y}$ give a worthwhile reduction in the remainder mean square? We must always remember that however well some F_i may predict values of y_i, the important thing statistically is the remainder mean square, which measures how much we don't know about y_i. Statements like 'the regression accounts for 86 per cent of the original variance' really mean 'the regression fails to account for 14 per cent of the variance'. Statisticians are trained to take a gloomy view.

We may try this analysis on the data given above. The mean of the eight values 31, 32, 29, 31, 30, 28, 30 and 29 is 30.0. The residuals are therefore 1, 2, -1, 1, 0, -2, 0 and -1. The sum of squares of the original values is 7212, the sum of squares of the remainders is 12 and so the analysis of variance is:

	Degrees of freedom	Sum of squares	Mean square
Due to fitting the mean	1	7200	7200.00
Residual	7	12	1.71
Total original	8	7212	901.50

Fitting the mean therefore reduces the mean square from 901.50 to 1.71. The corresponding variance ratio is 7200.00/1.71. Very often, as in this case, the overall mean is so obviously different from zero that we automatically fit the mean as a preliminary step before the analysis proper begins. In that case, the 'sum of squares due to fitting the mean' is called the 'correction for the mean', and we start the analysis with $N-1$ degrees of freedom. But there is no need to subtract the 'correction for the mean' in those cases where $\bar{y}$ does not differ significantly from zero; and occasionally it is not necessary to fit a mean, even though $\bar{y}$ obviously does differ from zero. We shall meet such a case in Chapter 3, namely a regression which must go through the origin.

The worked example shows why the variance ratio does not directly compare the original and remainder mean squares. The original sum of squares (7212) *includes* the remainder sum of squares (12). So the remainder mean square (1.71) is in some sense part of the original mean square (901.5); whereas 7200.00, the mean square due to fitting the mean, and 1.71, the remainder mean square, are distinct independent entities. The variance-ratio test can only compare two mean squares which are independent; neither sum of squares may 'contain' the other sum of squares, or any part of it. (It is always the sums of squares that add up in the analysis of variance, not the mean squares.) If the variance ratio is calculated to be less than one, it means that the original mean square was actually rather smaller than the residual mean square, so that the analysis has certainly not achieved a worthwhile reduction in the mean square. That is why published tables of variance ratio only consider values greater than one.

Very often, then, the overall mean obviously differs significantly from zero, and we are more interested in the difference between the means of two or more blocks of data. In that case, there is no doubt that fitting an overall mean greatly reduces the size of the residual mean square, and we want to know if a *further* reduction may be made by fitting separate means for each block. For example, is there any point in treating males and females separately? (There is clearly no sense in asking that question unless we can allocate each individual to its appropriate category, in this case male and female.) Suppose that the values of y for males are 31, 32, 29, 31, 30 and those for females 28, 30, 29. Fitting the overall mean will evidently give precisely the same analysis as before. The separate means for males and females are 30.6 and 29.0. The male residuals are therefore 0.4, 1.4, -1.6, 0.4 and -0.6, and the female residuals are -1.0, 1.0 and 0. The sum of the squares of all these residuals is 7.20. So, by fitting separate means for males and females, we get an analysis of variance:

	Degrees of freedom	Sum of squares	Mean square
Due to fitting male and female means	2	7204.80	3602.40
Remainder	6	7.20	1.20
Original	8	7212.00	901.50

The first line may be split:

	Degrees of freedom	Sum of squares	Mean square
Fitting overall mean (as before)	1	7200.00	7200.00
Extra for separate male, female means	1	4.80	4.80
Remainder	6	7.20	1.20
Original	8	7212.00	901.50

If we automatically correct for the overall mean before we start the analysis, we get the usual analysis of variance:

	Degrees of freedom	Sum of squares	Mean square
Males v. females, i.e. 'between sexes'	1	4.80	4.80
Remainder, i.e. 'within sexes'	6	7.20	1.20
Total (corrected for mean)	7	12.00	1.71

By distinguishing males from females, we reduce the residual mean square from 1.71 to 1.20. The variance ratio 4.80/1.20 is not significantly large, indicating that the reduction from 1.71 to 1.20 may well be fortuitous. Although the values of y may possibly be smaller, on average, for females than for males, the existing sample does not furnish enough evidence to make the distinction worthwhile. If the male and female means were identical, both would equal the overall sample mean, and so there would be *no* extra reduction in the residual sum of squares. Therefore, the mean square labelled 'males v. females' is indeed concerned with the average difference *between* males and females (Example 6.1).

Equally, the remainder sum of squares is made up of terms $(y_i - \bar{y})^2$, where y_i is a male (or female) observation and $\bar{y}$ is the corresponding male (female) mean. So the remainder sum of squares is made up from differences between males and males, or between females and females, i.e. from differences *within* the set of males and *within* the set of females. It is not affected by (i.e. it is independent of) the average difference between males and females. The whole analytic process of trying to reduce the residual mean square may be continued step by step, to see if any more detailed categorization improves the prediction of values of y. The smaller the residual mean square, the better we can predict y.

1.4. Weighting

So far we have implicitly assumed that all the values of y were measured with equal accuracy: each y has been given the same weight in the sum of squares. But sometimes we have to recognize that different ys have different accuracies, and then we wish to pay more attention to the most accurate values. Instead of a sum of squares $\sum(y_i - F_i)^2$ we use a weighted sum of squares $\sum w_i(y_i - F_i)^2$. Here w_i is the weight given to y_i. The usual unweighted sum of squares is a special case of the weighted sum of squares, when every $w_i = 1$. If F_i represents an overall mean m, the least-squares estimate of m (i.e. the estimate of m which minimizes the weighted sum of squares) is the weighted mean $\bar{y} = \sum w_i y_i / \sum w_i$. If every $w_i = 1$, this reduces to the usual unweighted expression $\bar{y} = \sum y_i / N$. If we use a weighted mean, we must also use a weighted analysis of variance— that is, an analysis of the weighted sums of squares. This shows how intimately the estimate of $\bar{y}$ is related to the analysis of variance. The unweighted mean does not minimize the weighted sum of squares, nor does the weighted mean minimize the unweighted sum of squares. To use a weighted mean in an unweighted analysis of variance (or vice versa) would be misleading, and therefore wrong.

Suppose that the variance V_i of the errors of measurement is not the same for every y_i in the sample. The most accurate value of $\bar{y}$ (i.e. that with the least error variance) is obtained by using a weight w_i equal to $1/V_i$. That means that an inaccurate value of y, with a large variance, is given a small weight. The weighted sum of squares then becomes $\sum(y_i - F_i)^2/V_i$. When dealing with continuous measurements, e.g. lengths or temperatures, we usually do not need to weight the analysis, because we can reasonably assume that V_i does not vary very much, in which case the weighted and unweighted analyses will return much the same biological answer. The point is discussed further in Chapter 5. Now suppose that y_i is a count which can take only discrete values $0, 1, 2 \ldots$, and suppose that the expected value of y_i is F_i. The Poisson distribution tells us to expect that V_i will equal the mean F_i. The weighted sum of squares then becomes $\sum(y_i - F_i)^2/F_i$, which is the usual expression for χ^2 (chi-squared) used in the analysis of data which take the form of counts. So the χ^2 analysis is a special case of the analysis of weighted sums of squares. Different statistical methods are closely related.

When we have to do a perplexing analysis, we can always in theory return to first principles and minimize the appropriate sum of squares (or maximize the appropriate likelihood—Chapter 7). But in a complicated case, the process of minimization may be just as perplexing. In practice, therefore, we work at a more superficial level, but using rules of statistical manipulation which are founded on least squares (or likelihood). These rules are examined in Chapters 2 and 3.

Notes

1. In the long run, the choice of the sum of squares is arbitrary (Chapter 7). It was originally adopted mainly because it led to easy methods of calculation. For example, we might choose to estimate m by minimizing the sum of the absolute values of the residuals, $\sum|y_i - m|$, instead. This function of m is not so easy to minimize because its slope changes discontinuously at each data point y_i. In fact it is minimized, not when $m =$ the arithmetic mean $\bar{y}$, but when $m =$ the median of the sample. But the convenience of the sum of squares goes rather deeper than mere simplicity of calculation. The quadratic is the only non-trivial continuous function of the data such that

$$f(a) + f(b) \text{ is equivalent to } f\left(\frac{a-b}{2}\right) + f\left(\frac{a+b}{2}\right),$$

on which fact the analysis of variance itself depends.

2. The difference between description and prediction is that we can always *describe* complete sets of existing data, but seek to *predict* single values that may arise in the future.

3. The term 'degrees of freedom' is perplexing. It is often used, as in this chapter, to mean 'the divisor needed to convert a sum of squares into a mean square which will estimate the population variance V'. For example, $\sum y_i^2$ has N degrees of freedom and $\sum(y_i - \bar{y})^2$ has $N-1$ degrees of freedom, because $\sum y_i^2/N$ will on

average equal V, if the population mean m is zero; and so will $\sum(y_i - \bar{y})^2/(N-1)$, whatever the value of m. The name 'degrees of freedom' arises in the following way. Once the value of $\bar{y}$ has been fixed, we could choose any values we liked for $N-1$ of the y_is, but then the remaining value of y would have to follow automatically, in order to give the right value of the sample mean $\bar{y}$. More precisely, the N values of $(y_i - \bar{y})$ must sum to zero, and so only $N-1$ of them are free to vary independently. Once $\bar{y}$ is fixed, the remainders $(y_i - \bar{y})$ have $N-1$ 'degrees of freedom'. It can be shown theoretically (e.g. Example 2.8) that the 'degrees of freedom' is indeed the right divisor to use to estimate V (but see Note 1, Chapter 4).

Examples 1

1.1. The sample values of y are 31, 32, 29, 31, 30, 28, 30 and 29. Calculate $\sum(y-m)^2$ for $m = 0, 10, 20, 30, 40$ and 50. Plot the values of $\sum(y-m)^2$ against m, to show that $\sum(y-m)^2$ is minimized when $m =$ the sample mean, 30. Using the plot, show how $\sum y^2$, i.e. $\sum(y-0)^2$, is split into $\sum(y-\bar{y})^2$ plus a part accounted for by fitting $\bar{y}$. Is there a value of m which reduces $\sum(y-m)^2$ to zero?

The next two examples require elementary algebra or calculus. They illustrate Chapter 1, but are not essential.

1.2. Show that, for *any* set of ys, $\sum(y-m)^2$ is minimized when $m = \bar{y}$.

1.3. Show that in a sample of N values of y with mean $\bar{y} = \sum y/N$,

$$\sum(y-\bar{y})^2 = \sum y^2 - N\bar{y}^2$$

and hence by rearrangement that

$$\underset{\substack{\text{original} \\ \text{sum of squares}}}{\sum y^2} = \underset{\substack{\text{remainder} \\ \text{sum of squares}}}{\sum(y-\bar{y})^2} + \underset{\substack{\text{sum of squares} \\ \text{due to fitting} \\ \text{the mean}}}{(\sum y)^2/N}$$

Verify this equation using the data of Example 1.1.

2 Means and variances

Chapter 1 showed how we use means to predict values of y, i.e. to reduce the sizes of the remainders. In this chapter the means will be treated as entities in their own right. The chapter is full of formulae, but the only difficult part is about two-way tables.

A single mean can describe and summarize a whole set of values of y. Its actual size will answer practical questions by telling us 'how much'. For example, given an average yield of wheat, we can decide how many acres to grow. Significance and practical importance are not the same thing. It can happen that although two means are significantly different, the difference is too small to bother about in practice. For instance, a new variety of wheat may consistently outyield an existing variety by 1 per cent; but that extra yield may not cover the cost of multiplying up seed stocks and marketing the new variety. Although a mean is really a predictor of individual values of y, we may also regard it as a descriptive 'statistic' in its own right.

One mean cannot entirely summarize a set of data. The two distributions shown in Fig. 2.1 are evidently quite different, although both have a mean of 30. To describe a distribution, we need to quote both its mean and its variance—or equivalently, its standard deviation. (If the distribution has some further peculiarity, e.g. if it is skew, still further description may be needed.) This takes us into the world of variances, standard deviations and standard errors. It is very important to grasp the elementary rules for handling these various quantities; otherwise you will be left floundering in a mire of Ns, $(N-1)$s and square roots. The rules are given in every textbook, but we shall re-examine them here.

2.1. Measures of dispersion

Suppose we wish to *describe* the distribution of y from a sample of N values. The average value $\bar{y}$ is $\sum y/N$. It is sometimes called a measure of location, because it locates the whereabouts of the values of y on the scale of measurement. We must also quote some measure of the spread, or dispersion, which will distinguish between the two cases in Fig. 2.1. In other words, we want some measure of the absolute size of $y_i - \bar{y}$.

Following the ideas of Chapter 1, we turn to the sum of squares $\sum(y_i - \bar{y})^2$. The sum of squares will not do as it stands, because we know that as the sample size N increases, so does the sum of squares. But the

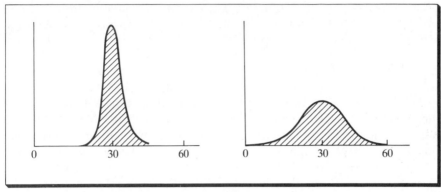

Fig. 2.1. Two distributions with mean 30.

mean square $\sum(y_i - \bar{y})^2/(N-1)$ remains approximately constant. It estimates the variance of single values of y. Thus the sample mean $\bar{y}$ estimates the overall average m of the whole population of values of y, and the sample mean square s^2 estimates the average of $(y_i - m)^2$, i.e. the variance V of individual values of y. So the mean square *estimates V*, just as $\bar{y}$ *estimates m*. The square root of the sample mean square is called the standard deviation, or root mean square. It is a measure of the average deviation of y from the population mean m. The mean, mean square, and standard deviation are all concerned with the individual values of y in the original population. They are, on average, unaffected by any change in sample size.

Usually, our sample of ys is a small sample from a large population. So the sample mean $\bar{y}$ is an inaccurate estimate of the population mean m. The bigger the sample size, the smaller the inaccuracy. We use a 'standard error' to estimate the size of the inaccuracy. The standard error is not concerned with the original population, but with the sample mean $\bar{y}$. So the standard error indicates the accuracy of the *sample mean*, and the standard deviation measures the variability of the individual values in the *population*. Regrettably, the word 'variance' is used in both these connections; we talk about 'the variance of a population' and 'the variance of a mean'. If s^2 is the value of the residual mean square—i.e. s^2 is an estimate of the population variance V—we estimate the variance of $\bar{y}$ as s^2/N. (This assumes that the residuals of successive values of y are uncorrelated, i.e. that there is no tendency for one residual to be specially large just because some other residual turns out large.) The *variance* of $\bar{y}$ is defined as the average value of $(\bar{y} - m)^2$ which we should get if we took a whole set of different samples, each of size N, and worked out a $\bar{y}$ for each sample. The *standard error* of $\bar{y}$ is the square root of the variance of $\bar{y}$, i.e. $\sqrt{(s^2/N)}$. It is a measure of the average deviation of $\bar{y}$ from m. If we doubled the sample size N, the residual mean square s^2 (i.e. the estimate of the variance of single y-observations) would remain approximately the same, but the estimated variance of $\bar{y}$, which is s^2/N, would be approximately halved—and so the standard error would be reduced by a factor of $\sqrt{2}$. This means that to double the accuracy of $\bar{y}$, we must multiply the sample size by 4; but to improve the accuracy of $\bar{y}$ tenfold, we must multiply N by 100. Such is life.

There are two ways of joining the last two paragraphs together. First, a single value of y might be regarded as a sample of size $N = 1$. Thus the standard deviation (which refers to any one original observation) is the standard error of a single value of y (regarding that value as an estimate of the population mean m). Second, we divide the corrected sum of squares $\sum(y_i - \bar{y})^2$ by the degrees of freedom $(N-1)$ to get s^2, the remainder mean square. The value of s^2 will come out much the same, however large the sample size N. To estimate the variance of the sample mean $\bar{y}$, we further divide s^2 by N, because $\bar{y}$ is the mean of N values of y. This variance of $\bar{y}$ (i.e. s^2/N) decreases as N increases, because $\bar{y}$ becomes more accurate as the sample size gets bigger. So in a formula like 'the variance of $\bar{y}$ is $\sum(y_i - \bar{y})^2/N(N-1)$', the N and $N-1$ are serving quite different purposes. The $N-1$ is the degrees of freedom needed to reduce the sum of squares to a mean square which estimates the variance of single values of y, and the N is there because $\bar{y}$ is the mean of N different values of y.

2.2. Manipulation techniques

There are some simple rules for manipulating means and variances. If the variance of y is V and a is a constant, the variance of ay is $a^2 V$. If the variance of y_1 is V_1 and the variance of y_2 is V_2, the variance of $y_1 + y_2$ is $V_1 + V_2$, provided that y_1 and y_2 are uncorrelated. These rules hold true whatever the ys may be—original observations, means, totals, or regression coefficients. We may combine the two rules by saying that the variance of $a_1 y_1 + a_2 y_2 + \cdots$ is $a_1^2 V_1 + a_2^2 V_2 + \cdots$, provided that the ys are uncorrelated.[1] This formula may be used to work out any particular case that may arise. For example, if we put $a_1 = 1$ and $a_2 = -1$, we obtain the variance of $y_1 - y_2$ as $V_1 + V_2$, which is also the variance of $y_1 + y_2$. To get the variance of the sum, or of the difference, of two numbers we add together their respective variances. We can use this same rule to find the variance of a mean. If we have N different ys, and if we put $a_1, a_2, \ldots, a_N$ all equal to $1/N$, we find that the variance of $\bar{y}$ (i.e. the variance of $y_1/N + y_2/N + \cdots + y_N/N$) is $V_1/N^2 + V_2/N^2 + \cdots + V_N/N^2$. If the Vs are all the same, the variance of y is therefore NV/N^2, or V/N. So the variance of the mean is $1/N$ times the variance of a single observation, as stated before. The condition that the ys shall be uncorrelated is very important.[2] If the ys are correlated, extra terms called covariances must accordingly be added to the formula. When we use the formula in practice, the estimates of V are usually calculated from *remainder* mean squares, and so it is the *remainders* which have to be uncorrelated: this is usually the case, since the original ys have been measured independently from independent biological materials, e.g. different animals. There are also rules for finding the variance of $y_1 \times y_2$, of y^2, of y_1/y_2 or of any other non-additive combination of ys, but it is wise to seek expert advice when such problems arise.

Suppose that we wish to compare two means $\bar{y}_1$ and $\bar{y}_2$, which have variances V_1 and V_2. The variance of $\bar{y}_1 - \bar{y}_2$ is $V_1 + V_2$, provided that the $\bar{y}$s are uncorrelated. So it is the *variances* that add together. We have already met this additive property of variances in another guise in Chapter 1, where we divided the total sum of squares into a part accounted for by the mean plus a remainder part. Because of their additive properties, we work primarily with variances. For example, if $\bar{y}_1$ has standard error (S.E.) s_1 and $\bar{y}_2$ has S.E. s_2, the variance of $\bar{y}_1 - 2\bar{y}_2$ is $s_1^2 + 4s_2^2$, i.e. we work in terms of the variances s_1^2 and s_2^2. Only at the very end of the calculation do we find the standard error, in this case $\sqrt{(s_1^2 + 4s_2^2)}$, by taking the square root of the variance.

In the example considered in Chapter 1, the residual mean square (after fitting the means for male and female) is 1.20. This is an estimate of the variance of individual measurements, applying equally to males and females. (If we had suspected that males and females have different variances, we should accordingly have estimated those variances separately.) The mean for males is 30.6, and since it is the mean of five observations, its variance is estimated as 1.20/5. Similarly the variance of the female mean (29.0) is 1.20/3. The difference between the means is $30.6 - 29.0 = 1.6$, and its variance is the sum of the variances of the means, $1.20/5 + 1.20/3 = 0.64$. This is in fact the expression $s^2/N_1 + s^2/N_2$ or $s^2(1/N_1 + 1/N_2)$ which appears in the formula for t to test the difference between the two means. The standard error of the difference is the square root of the variance, $\sqrt{0.64}$ or 0.80. Thus the difference between the male and female means, 1.6, is exactly twice its standard error 0.80. The basic steps are always the same:

(1) calculate the residual mean square;
(2) use the rules given above to calculate the variance of some interesting combination of the original ys—in this case, the difference between the two means;
(3) take the square root of that variance to find the standard error.

Chapter 6 will consider how to use the S.E. to test significance.

2.3. One-way table

Suppose we have done an experiment involving a single set of different
treatments, so that to every y there is assigned one, and only one,
treatment. We have just considered an example with two treatments, viz.
male and female. But now suppose that there are several treatments. The
data form a 'one-way table', because they are classified into a single set
of treatments and so could be written in a single row and divided up
according to those treatments. For example, the data in Example 2.2
could be written:

1.07 1.20 0.98 1.07 | 1.20 1.31 1.36 1.27 | 1.35 1.41 1.40 4.37 1.36 1.39.

The analysis follows the principles stated in Chapter 1. The overall mean
is subtracted from each observation, and the sum of squares of the
remainders is the 'original sum of squares corrected for the mean'. Then
individual treatment means are used to calculate a new set of residuals
whose sum of squares is the 'residual s.s.' The difference between these
two sums of squares is the 'sum of squares due to fitting separate
treatment means', or 'treatments s.s.' for short. It is concerned, as before,
with the *differences* among a set of treatments.

Now we wish to compare the set of treatment means. If we choose the
biggest and smallest means, just because they *are* the biggest and
smallest, they will quite possibly appear significantly different, as judged
by a *t*-test—but that conclusion may well be spurious. The biggest and
smallest numbers chosen from a set of random numbers may also appear
significantly different, until we remember that they have been specially
chosen from the whole set of numbers. There is therefore a danger of
drawing unjustified conclusions when comparing a set of treatment
means. True, there are special multiple-range tests (Chapter 6) designed
to avoid that danger. But the examination of a set of treatment means
requires not only statistical tests, but also common sense and biological
gumption. We have to bear in mind the biological story as well as the
statistical significance. No amount of statistical manipulation can relieve
us of responsibility for the biological conclusions which we draw from
the data.

It is possible to make many comparisons between a set of treatment
means. Some of those comparisons will answer questions which you
would have wanted to ask *before* you collected the data. It is always valid
to make an appropriate significance test on any such comparison. But
other comparisons are interesting, not *a priori*, but because the data
themselves suggest some unanticipated effect. There is then a danger of
mistaking chance differences for real biological effects; but at the other
extreme, it is possible to ignore unexpected differences which the data

insist are really there. The following rule, although not foolproof, makes a useful guide. Suppose the variance ratio (treatments mean square/ residual m.s.) is not significant, suggesting that there are no differences among the set of treatments as a whole. There may still be one or two individual differences which have been submerged, when ploughed in with the other treatment means; the differences, although really there, not being big enough to significantly inflate the between-treatments mean square. In that case, *a priori* comparisons decided on before the experiment was done may be made with confidence; but it would be very dangerous to rely on apparent differences which are noticed only when the data are inspected.

On the other hand, if the variance ratio indicates that some real differences exist between the treatments, we may not only make *a priori* comparisons, but also have more confidence in unanticipated differences. It is still wise not to put too much trust in such unexpected conclusions, but merely to regard them as interesting suggestions for further experimentation. Now suppose the treatments can be arranged in order *a priori*, e.g. they might represent increasing doses of a fertilizer. Then the treatment means may show a genuine pattern, e.g. they may increase steadily as the fertilizer dose increases, even though the differences between consecutive means may not be significant.

There are innumerable different comparisons which can be made between a set of means, but those comparisons are not all independent of each other. Suppose that in some experiment, the mean of the first treatment has by chance come out rather too big. Then the value of $\bar{y}_1 - \bar{y}_2$ will be too big, and so will $\bar{y}_1 - \bar{y}_3$. If the difference $\bar{y}_1 - \bar{y}_2$ is mistakenly judged to be significant, it is quite likely that $\bar{y}_1 - \bar{y}_3$ will also appear to be significant. That does not mean that we may not test both $\bar{y}_1 - \bar{y}_2$ and $\bar{y}_1 - \bar{y}_3$ if we want to, but we should remember that the two comparisons are not independent. Fortunately it is often the overall pattern of the whole set of means that is interesting, rather than any particular comparison between individual means. It is usually quite sufficient to calculate the set of treatment means with their standard errors, and inspect them carefully to draw the appropriate biological conclusions. That is why it is so important to grasp the rules discussed earlier in this chapter for handling means, standard deviations, variances and standard errors.

Suppose that there are real differences between treatments, and that an overall mean is required for the whole set of treatments. If one treatment is represented by more *y*-values than another treatment, the overall grand mean will favour the first treatment. To avoid bias it may be advisable to take the unweighted mean of the individual treatment means (Example 2.1). Such a mean would be less precise (i.e. have a bigger standard error) than the overall mean, but would avoid the possibility of bias. The difficulty does not arise if every treatment has the same number of y-observations.

2.4. Nested one-way table

A 'nested', or hierarchical, classification is a simple extension of a one-way table. Some or all of the treatments in the one-way table are sub-divided into sub-treatments. For example, the ordinary Linnean nomenclature is hierarchical: the generic name is the treatment and the specific name is the sub-treatment. Each species (sub-treatment) belongs to only one genus (treatment). Example 2.3 shows the detailed analysis of a nested table. You would be wise to work through Examples 2.1–2.3 before attempting the complexities of a two-way table.

2.5. Two-way table

Until now we have considered only a single set of treatments. For example, an animal may belong to one species or another, but not to two at once. Now suppose that there are two sets of treatments superimposed on each other. The y-values are classified simultaneously into 'rows' and 'columns'. Each row represents one particular treatment from the first set of treatments, and each column represents a treatment from the second set. It is no longer possible to display the data in a single line and still show the structure of the treatments: instead, a two-dimensional array is needed. If our dogs, besides being male and female, may be of different breeds, we might have a two-way table as follows:

	Alsatian	Borzoi	Collie	Dingo
Male				
Female				

The two sets of treatments cut across each other, because a male can belong to any breed and an Alsatian may be of either sex. It is this circumstance which distinguishes a two-way table from a nested one-way table. In the one-way table, once the species is known, the genus follows automatically; in the two-way table, both breed and sex must be specified.

In a two-way table, each particular combination of a row and a column is called a 'cell'. Each value of y is assigned to its appropriate cell.

2.5.1. Orthogonal two-way table

If the number of observations per cell is the same in every cell, the table is 'orthogonal' (but tables may also be orthogonal in some other circumstances.) Orthogonality has nothing to do with the actual values of y, but with the pattern of frequencies in the cells.

The analysis begins by forgetting about the row and column classification. Instead it regards each cell as a separate treatment in its own right:

$$y_{ij} = \text{appropriate cell mean} + \text{remainder.} \qquad (2.1)$$

This gives a simple one-way analysis which separates the 'between cells' and 'remainder' sums of squares. The analysis then proceeds to recognize the row and column classifications:

$$y_{ij} = m + a_i + b_j + \text{interaction} + \text{remainder.} \qquad (2.2)$$

Here the remainder is the same as in eqn (2.1), and the four terms m, a_i, b_j and interaction dissect the cell mean in eqn (2.1)—the point being that the *same* row constant a_i is used for every cell in the ith row, and the *same* b_j for every cell in the jth column. Consequently, if there is a genuine pattern of row effects, i.e. if the values of y in one row tend to be consistently larger than those in another row (irrespective of the particular column), the difference will show up in the corresponding as— and similarly for columns. Equation (2.2) means that any value of y in the cell formed by row i and column j is the sum of the following:

1. An overall constant m which applies to all values of y in the table. If the table is orthogonal, m equals the overall mean of y; otherwise, m is corrected to allow for the fact that some rows and columns would be represented in the overall mean more frequently than others.

2. A row constant a_i which applies to all values of y in row i. If any arbitrary constant were added to m and the same constant subtracted from every a_i, eqn (2.2) as it stands would be unaffected. In other words, some of the constants in eqn (2.2) as it stands are superfluous. Another way of saying the same thing is that once the overall mean m has been fitted, there are only $r - 1$ degrees of freedom between the r row constants a_i. To avoid the ambiguity, we usually require quite arbitrarily that the values of a_i shall add up to zero, i.e. that the average value of all the as is zero. Then m takes care of the overall average, while the as are concerned with *differences* between rows.

3. A column constant b_j which applies to all the values of y in column j. The average value of all the bs is similarly set equal to the arbitrary

value zero. The analysis is symmetrical for rows and columns, i.e. interchanging rows and columns will produce exactly the same analysis, but with the row and column constants interchanged.

4. An interaction which applies to all values of y in the cell (i,j). The term 'interaction' is used here in a purely statistical sense. If rows and columns interact, it means that the difference between two rows is not always the same, but depends to some extent on the particular choice of column; and vice versa. An interaction may sometimes be regarded as a statistical parameter in its own right (like m, a_i and b_j) but it is usually preferable to regard it as 'a bit left over after the additive terms m, a_i and b_j have done their best'. The interactions are really a measure of the failure of the purely additive model

$$y_{ij} = m + a_i + b_j$$

to describe the cell means completely. If interactions occur, that additive model is not a perfect way of viewing the situation. The interaction is in fact estimated as the difference between the cell mean and $(m + a_i + b_j)$—compare eqns (2.1) and (2.2). Therefore, a large interaction indicates that the individual cell means do not fit the additive pattern perfectly—or in other words, we could predict individual values of y more accurately by using the cell mean than by using the row and column constants. So why bother with rows and columns?—because they give greater generality.

It might be possible to find some transformation of y to a new scale of measurement (Chapter 5) which eliminates interactions, so that the row and column treatments act together in perfectly additive fashion on the new scale. That would be advantageous from a statistical point of view, it being much easier to think about the situation when there are no interactions.

5. A remainder which applies only to the particular value of y, and which is included because values of y in the same cell will not all be the same. If there is no more than one value of y in any cell, there will be no remainder variation. Such a situation sometimes defeats the object of the analysis, and may be overcome by reducing the number of rows and columns by grouping together similar rows or columns.

The *analysis* of orthogonal tables is straightforward. It indicates how much the mean square is reduced by fitting a constant a_i for each row, how much it is reduced by fitting column constants b_j, and how much it is further reduced by fitting interactions. The rows account for so much of the original sum of squares, the columns for so much, and the interactions for so much, leaving so much as residual. The *interpretation* of the results is more complicated. Some people say that the interpretation will vary according to whether we consider rows and columns to have fixed or random effects—the terms are explained in Chapter 4. Other people think that the distinction between fixed and random is inappropriate or unnecessary. It is certain that in practice, the distinction makes no great difference to the answer. If it did, we could not believe that answer anyway. The following method of interpretation may safely be used in all cases.

Any mean square in the analysis of variance may be compared with the remainder mean square. The appropriate significance test is the variance ratio or F-test. If we test the rows mean square against the remainder mean square, we are asking 'Are there apparent differences between rows?' But suppose there are genuine interactions between rows and columns. That means that the effects of rows are not constant, but vary from column to column, and vice versa. Usually the average row and column effects are bigger than the interactions, in which case it is still useful to think in terms of average row and column effects while recognizing that they interact to some extent. But occasionally the row and column effects are no larger than the interactions. In that case the additive analysis based on eqn (2.2) is not a helpful way of viewing the situation, since the difference between two rows depends entirely on the particular column involved. Instead of thinking in terms of average row and column effects, we may just as well treat each row–column cell as an individual treatment—unless some more appropriate non-additive model can be used instead of eqn (2.2) to specify how the rows and columns act together.

The first step in the interpretation of the analysis of variance, therefore, is to test the interactions mean square against the remainder mean square. If those two mean squares are about the same size, either may be used as an estimate of residual variance. If degrees of freedom are few, a combined mean square may be calculated as follows. Add together the interactions and remainder sums of squares, and divide by the sum of the corresponding degrees of freedom. The result must lie somewhere between the interactions and remainder mean squares. This procedure assumes that there are *no* genuine interactions, so that the interactions mean square is just another independent estimate of the residual variance. On the other hand, if there *are* genuine interactions, we shall need to use the interactions mean square as a residual, because we want to ask, not 'Are there average differences between rows or columns?' but 'Are the row and column effects more important than their interactions?' So we shall compare the rows or columns mean square with the interactions mean square, rather than with the within-cell residual. (Those who know about components of variance will realize that this may be a rough and ready procedure, but it is usually good enough in practice.) And if there is no residual, we have to use the interactions mean square anyway.

The method of interpretation just described is based on one simple principle. To examine the reality of row differences, the rows mean square must be compared with some other mean square which estimates what the rows mean square would be if there were no true row effects. That principle applies, not just to two-way tables, but in all situations. An important point in experimental design is to ensure that an appropriate residual mean square can indeed be estimated (Chapter 8). If we ask 'Do row or column effects exist, regardless of their interactions?' the appropriate variance ratio is the rows (or columns) mean square/the within-cell residual mean square. But if, as suggested, it is better to ask 'Are row and column effects more important than their interactions?' the variance ratio becomes the rows (columns) mean square/the interactions mean square. It can happen, but very rarely, that interactions occur between non-existent row and column effects. You may like to consider the implications of that situation.

Before attempting the next section, you should work through Example 2.4, which analyses an orthogonal two-way table. The next section is a complicated example of the simple principles of Chapter 1.

2.5.2. Non-orthogonal two-way table

The analysis of a non-orthogonal two-way table is more complex than the analysis of an orthogonal table, and its interpretation is more problematical. Wherever possible, experiments and samples should be designed to avoid non-orthogonality. But sometimes the biologist cannot specify in advance how many observations will be collected in each cell.

If the table is non-orthogonal, it is not possible to separate out the effects of rows and columns completely. It is easy to see why not. The average of y for one row will include the effects of some particular assortment of columns, while another row will be associated with a different combination. Therefore, however intensively we analyse, we can never be entirely sure that an apparent difference between rows is not to some extent a concealed effect of columns, transmitted via the uneven pattern of cell frequencies. And conversely, apparent differences between columns may in part be second-hand effects of rows. This explains why, in a non-orthogonal table, the sums of squares for rows and columns no longer add up nicely, as they did for the orthogonal table. The straightforward sums of squares for rows and columns are not independent. The logical pattern of the analysis (Fig. 2.2) is adjusted accordingly. We proceed as follows.

At each step, the analysis includes an overall mean m, so that as before the row constants a_i add up to zero, and so do the column constants. First we fit row and column constants simultaneously by least squares, absorbing a sum of squares (1) for rows and columns together. This step

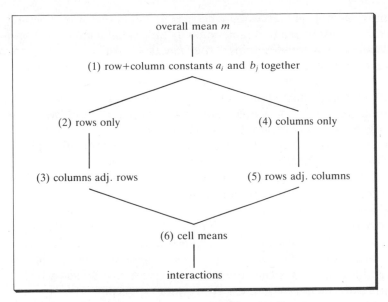

Fig. 2.2. Logical scheme of analysis of a non-orthogonal two-way table.

involves heavy calculation which cannot easily be done by desk calculator, but is child's play to a computer. Then we split the sum of squares for rows and columns in two different ways (Fig. 2.2). We start by fitting row constants only, giving the usual straightforward one-way analysis (between rows and within rows) which absorbs a sum of squares (2) for rows. The difference between this sum of squares (2) for rows and the sum of squares (1) for rows and columns is the *extra* sum of squares (3) absorbed by columns, after the row constants have been fitted. It is called 'sum of squares for columns adjusted for rows' or 'columns adj. rows' for short. It must represent effects of columns only, since all the effects of rows have already been accounted for. But it need not show the *whole* effect of columns, because as mentioned above, the rows sum of squares (2) may include some indirect effects of columns which show up in the row means.

In this way we have split the sum of squares for rows and columns together (1), into a sum of squares for rows (2) plus a sum of squares for columns adj. rows (3). We now start again, and split the same sum of squares for rows and columns together (1), into a sum of squares for columns (4) plus a sum of squares for rows adj. columns (5). This double split (Fig. 2.2) is the best we can do to divide the variance of y into a part for rows and a part for columns. If the table were orthogonal, the split would be clean and unambiguous. The sum of squares for columns would be the same as the sum of squares for columns adj. rows, and the sum of squares for rows would equal the sum of squares for rows adj.

columns. Orthogonal tables may be recognized in that way. The fact that the split cannot be made unambiguously unless the table is orthogonal is the main reason for designing experiments to avoid non-orthogonality wherever possible. Another reason is to get as much information as possible for our money.

The analysis ends by calculating the interactions sum of squares, obtained (as in an orthogonal table) by subtracting the rows + columns sum of squares (1) from the between-cells sum of squares (6). The residual is obtained as usual from the variation within cells. The whole procedure is illustrated step-by-step in Example 3.8.

Once the analysis is complete, we have to interpret what it means. Are there genuine effects of rows and columns? If the mean squares for both 'rows' and 'rows adj. columns' are significantly large, it means that there are real row effects. But it can happen that the 'rows' mean square is large, but the 'rows adj. columns' mean square is not. (That can't happen in an orthogonal table, where those two mean squares are necessarily identical.) That means that there are apparent row effects which *might* be indirect effects of columns. We cannot be certain. All we can say is that the observed row effects *can* be explained away as the indirect effects of columns. It is unusual, although not impossible, for very large row effects to be indirect effects of columns. If neither 'rows' nor 'rows adj. columns' is significant, the implication is that there are no true row effects at all, so far as the data go. It sometimes happens that a mean square is large enough to suggest that there may be some effect, but not large enough to insist that the effect is really there. In that case, it helps to look at the pattern of values of the row constants a_i to see if they make biological sense. As in any other analysis, mechanical use of significance tests cannot absolve the biologist from responsibility for the interpretation.

Orthogonal tables may be analysed if necessary by desk calculator. But the correct analysis of a non-orthogonal table is very laborious (except when there are only two rows or two columns—Example 2.10). Some computer programs short-cut the non-orthogonal analysis. They calculate a mean for each cell and then analyse those means as if the table were orthogonal, i.e. had the same number of observations in every cell. Such programs require that there must be at least one observation in every cell so that a mean can be calculated, which is not a requirement for the regular analysis described above. This short-cut procedure certainly eliminates the ambiguity between rows and columns, but only at a price. It throws away information about the row and column effects, by assuming that all cell means are equally accurate, when in fact some are more accurate than others and should be weighted accordingly—as they are in the regular analysis. If the numbers of observations per cell are more or less the same, the loss of information is slight—but so is the ambiguity of the regular analysis. If the non-orthogonality is severe, the short-cut method in effect throws away a large part of the data merely to get a simple analysis. That might be reasonable if the data are cheap to collect—but real statisticians are made of sterner stuff.

Sometimes a two-way table is designed to be orthogonal, but one or two values are accidentally missing. Before the advent of computers, special missing-value techniques were used to analyse such tables as if they were truly orthogonal. Such techniques, if used properly, must necessarily give the same answers as the appropriate non-orthogonal analysis. These days, we simply do the non-orthogonal analysis by computer. Since the table is very nearly orthogonal, there is very little ambiguity in its interpretation.

The difficulties of analysis and interpretation of non-orthogonal two-way tables have made this discussion unduly long, considering the relative rarity of such tables. These difficulties should at least convince you that non-orthogonal tables are best avoided wherever possible. Three-way tables (or those with more than three ways), whether orthogonal or not, are analysed following the same principles as we have used for two-way tables. The interpretation of a seriously non-orthogonal three-way table can be a nightmare of ambiguity.

2.6. Presentation of results

The presentation of results is largely a matter of personal preference—or of argument with editors and referees who sometimes insist on unnecessary statistical detail. Sprent (1970) castigates 'misguided editors who think that all numerical results can be made respectable by quoting significant differences or significance levels—often denoted by * or ** or ***, a symbolism more appropriate to a hotel guide-book than a serious scientific paper'. There is, regrettably, often a sharp distinction between statistical analysis performed to find an answer, and that done for publication. Elaborate statistical analyses are presented to prove points which are perfectly obvious. That is not to say that statistical analyses should not be included in a biological paper, but that they should appear only when necessary and relevant. There are two considerations here. First, it is important to supply the reader with enough detail so that he can, if he thinks fit, draw different conclusions from the same data. But on the other hand, the more tables of numbers there are in a paper, the fewer people will read it.

Very often, a table of treatment means and standard errors is all that is necessary. If all the means have the same standard error, it need be quoted only once; and if the means all have approximately the same standard error, an average is usually good enough. But if the means have very different accuracies, you should indicate the accuracy of each mean, either by quoting its own standard error or by showing the number of observations on which that mean is based—together with the standard deviation or some other estimate of overall variability.

In biology, it is rarely worth quoting more than three significant figures in any mean. If someone says 'a duck lays on average 4.603 eggs', the final figure 3 is almost certainly worthless. It is useful to quote one more decimal place in the standard error than in the mean itself, because that helps to preserve accuracy in subsequent calculations, e.g. of confidence limits. Given the means and standard errors, you can deduce all sorts of calculations, e.g. analysis of variance, significance tests, confidence limits: but it is not possible to calculate backwards from a significance probability to a standard error. It should be clearly stated that quoted standard errors *are* standard errors, rather than standard deviations or confidence intervals. The enthusiastic quoting and starring of significance probabilities, although very popular,[3] is inadvisable because it betrays a misunderstanding of the role of a significance test (Chapter 6). It is always easier to look at a picture, graph or histogram than to examine a table of numbers. So it's a good idea to present results pictorially wherever suitable. But information cannot be stated so accurately on a graph as in numerical form. And it can cost even more to print a figure than to print a table of numbers.

Notes

1. *Proofs of formulae* are provided for the sceptical, but not essential for practical understanding. The symbol E means 'the expected, or average, value of'. By definition $E(y) = m$ and $E(y - m)^2 = V$. That is what we mean by 'the mean m' and 'the variance V'. Now suppose that every value of y is multiplied by a constant a. The average value of y is multiplied by the same constant, i.e. $E(ay) = am$. The variance of (ay) is, by definition, the average value of $(ay - \text{its mean})^2$, i.e. $E(ay - am)^2 = a^2 E(y - m)^2 = a^2 V$.

Next consider y_1 and y_2 with means m_1, m_2 and variance V_1, V_2 respectively, so that $a_1 y_1$ and $a_2 y_2$ (where a_1, a_2 are constants) have means $a_1 m_1$, $a_2 m_2$ and variances $a_1^2 V_1$, $a_2^2 V_2$. The mean of $(a_1 y_1 + a_2 y_2)$ is then $(a_1 m_1 + a_2 m_2)$. The variance of $(a_1 y_1 + a_2 y_2)$ is therefore

$$E(a_1 y_1 + a_2 y_2 - a_1 m_1 - a_2 m_2)^2$$
$$= E[a_1(y_1 - m_1) + a_2(y_2 - m_2)]^2$$
$$= a_1^2 E(y_1 - m_1)^2 + 2a_1 a_2 E(y_1 - m_1)(y_2 - m_2) + a_2^2 E(y_2 - m_2)^2$$
$$= a_1^2 V_1 + a_2^2 V_2 \quad \text{provided that } E(y_1 - m_1)(y_2 - m_2) = 0.$$

The latter condition is met provided that y_1 and y_2 are uncorrelated so that the residual $(y_1 - m_1)$ is equally likely to be positive or negative, whatever the value of $(y_2 - m_2)$ may be. The same argument may be repeated to prove that since the variance of $a_1y_1 + a_2y_2$ is $a_1^2V_1 + a_2^2V_2$, the variance of $a_1y_1 + a_2y_2 + a_3y_3$ is $a_1^2V_1 + a_2^2V_2 + a_3^2V_3$, provided that the ys are all uncorrelated. And so on.

Now consider the weighted mean

$$(w_1y_1 + w_2y_2 + \cdots + w_Ny_N)/(w_1 + w_2 + \cdots + w_N)$$

We want to find the values of w_1, w_2 which will give this weighted mean the smallest possible variance, i.e. make the weighted mean as accurate as possible. The coefficient a_1 of y_1 is $w_1/(w_1 + w_2 + \cdots + w_N)$ and similarly for the other as. The variance of the weighted mean, $a_1^2V_1 + a_2^2V_2$ etc., is then

$$(w_1^2V_1 + w_2^2V_2 + \cdots + w_N^2V_N)/(w_1 + w_2 + \cdots + w_N)^2.$$

To find the values of w_1, w_2 etc. which minimize this variance, we set the differential with respect to w_1 equal to zero, i.e.

$$2w_1V_1/(w_1 + w_2 + \cdots + w_N)^2$$
$$- 2(w_1^2V_1 + w_2^2V_2 + \cdots + w_N^2V_N)/(w_1 + w_2 + \cdots + w_N)^3 = 0.$$

There is a similar equation for each of w_2, w_3, ..., w_N, and the set of N such equations (one for each w) together specify the required values of the ws. The second term of the equation is the same in every case, and therefore the first terms must be equal too, i.e. w_1V_1, w_2V_2, ..., w_NV_N must all be equal. Therefore, the most accurate weighted mean is obtained when $w_i = 1/V_i$, i.e. each y_i is weighted inversely by its variance. It makes no difference if we multiply each weight by a constant a, making $w_i = a/V_i$, since the constant a cancels out when the weighted mean is calculated. Therefore, when every V_i is the same and equal to V, we may set every w_i equal to $1/V$ or equivalently, every $w_i = 1$. The weighted mean then reduces to the ordinary arithmetic mean, and its variance to V/N (Example 2.9), always provided that the ys are uncorrelated.

2. The rule for finding the variance of $a_1y_1 + a_2y_2 + \cdots$ works only if the ys are uncorrelated. Otherwise, if the variance of y were V, the variance of $(y+y)/2$ would be $V/2$, i.e. by doubling y and halving again we could reduce its variance! This ridiculous result, arising because y is not uncorrelated with itself, shows how important is the requirement that the ys must be uncorrelated.

Finite population. A special case arises when samples are drawn from a population of finite size. The formulae for standard errors given in the chapter assume that the population is infinite, or in practice that the sample contains much less than 1 per cent of the total population at risk of being sampled. For example, the population of 'all possible Australian beer-drinkers' in Example 2.5 is so large that it is effectively infinite.

But now suppose that the population is small, and that the sample contains a fraction f (the sampling fraction) of the total population. For an infinite population, f must be zero. If $f = 1$, the sample contains the whole population; the population still has a standard deviation, but the sample mean is precisely the population mean, and so its standard error is zero. In this situation, the residuals in the sample are not independent—they must add up to zero, to make the sample mean equal to the population mean. In the general case, the variance of the sample mean is $V(1 - f)/N$. When $f = 1$ this is zero as required, and in the usual case of an infinite population, $f = 0$ and the variance of the sample mean becomes V/N as usual.

3. The deplorable use of stars to denote significance arose by accident. The eminent statistician Frank Yates (1937) indicated in a footnote that the value of a quoted statistic exceeded the 5 per cent significance level. He used a star to refer to the footnote. A subsequent value on the same page exceeded the 1 per cent significance level, necessitating a second footnote which received two stars to distinguish it from the first footnote. No correspondence between the number of stars and the level of significance probability was intended! Dr Yates would be the first to decry the mechanical use of significance levels, which the star system embodies.

Examples 2

As far as possible, Examples should be worked using a computer.

2.1. To be done twice, once by desk calculator and once by computer. A sample of eight Loch Ness monsters is found to contain five males and three females. Their lengths in feet are

Males: 39, 40, 37, 39, 38;   Females: 28, 30, 29.

Do a one-way analysis of variance. Find the mean of each sex, the overall sample mean, the average of the male and female means, and the difference between the male and female means. Estimate the variances of those five quantities.

2.2. Average daily increases in length (cm) of three lots of pigs were

Lot 1 Large White	Lot 2 Landrace	Lot 3 Large White × Landrace
1.07	1.20	1.35
1.20	1.31	1.41
0.98	1.36	1.40
1.07	1.27	4.37
		1.36
		1.39

Calculate means $\bar{y}_1$, $\bar{y}_2$ and $\bar{y}_3$ (with standard errors) for each lot. Calculate

$$\bar{y}_2 - \bar{y}_2, \quad \bar{y}_3 - \bar{y}_2, \quad \text{and} \quad \bar{y}_3 - (\bar{y}_1 + \bar{y}_2)/2$$

together with their standard errors. What is the point of making these particular comparisons?

2.3. A nested or hierarchical classification. A set of sparrows' nests each contained three chicks. Extra food was given to the parents of some nests, not to others. The chicks were weighed at ten-day intervals. Gains in weight (g) of each chick were

Treatment	Nest number	Weight gains
Control—no extra food	1	1.6, 2.0, 2.0
	2	0.8, 0.7, 0.3
	3	1.2, 1.9, 1.5
	4	0.5, 0.8, 1.4
	5	1.3, 0.2, 1.0

Treatment	Nest number	Weight gains
Extra food	6	1.6, 1.7, 2.1
	7	2.2, 2.0, 2.2
	8	2.4, 1.6, 2.2
	9	1.3, 1.9, 1.9
	10	2.0, 2.4, 1.7

Compare the average weight gains of chicks in the two treatments.

2.4. Sprinting speeds (ft/s) of various animals were

	Cheetah	Greyhound	Kangaroo
Males	56, 52, 55	37, 33, 40	44, 48, 47
Females	53, 50, 51	38, 42, 41	39, 41, 36

Analyse and interpret this two-way table.

2.5. To be done only by computer, using a program for two-way non-orthogonal tables. Quantities of beer (pints) drunk in a contest by various individuals were

Australians (male)	12.3, 14.0, 11.5, 9.7, 11.4, 10.8
British (male)	10.4, 12.5
British (female)	7.8, 7.0, 5.9, 6.0
Germans (male)	10.0, 11.0, 12.0
Germans (female)	6.0, 7.0, 7.0

(a) Do a one-way analysis of nationalities, pooling sexes; what is your conclusion?
(b) Do a two-way analysis of both nationality and sex; does this analysis modify the previous conclusion?

2.6. What would your interpretation be if, in a non-orthogonal two-way table, the 'columns' and 'rows adj. columns' mean squares were both large, but the 'rows' mean square was not?

2.7. Show that, in a two-way table, the 'sum of squares for rows and columns together', i.e. the sum of squares accounted for by fitting row and column constants simultaneously, cannot be less than the 'sum of squares for rows', even though the actual row constants may be different in the two cases. (*Hint*: what principle is used to estimate the row and column constants?) Hence show that the sum of squares for 'columns adj. rows' cannot be negative.

2.8. Show that if y has mean m and variance V, then the average value of y^2 is $m^2 + V$. From Example 1.3, $\sum(y - \bar{y})^2 = \sum y^2 - N\bar{y}^2$. Show that the average value of $\sum(y - \bar{y})^2$ is $(N-1)V$. Therefore, if the sum of squares $\sum(y - \bar{y})^2$ is divided by its degrees of freedom $N-1$, the mean square estimates V, whatever the value of m may be. What happens if $N=1$? Show that the original mean square $\sum y^2/N$ over-estimates V unless $m = 0$.

2.9. *Weighted analysis.* If the variance of y is obviously not constant, and the trouble cannot be dealt with by transformation (Chapter 5), a weighted analysis may be necessary. If the variance of y_i is V_i, the weight w_i given to y_i is $1/V_i$ (Note 1). The weighted mean $\bar{y}$ is then $\sum w_i y_i / \sum w_i$. Using the rule given in this chapter, prove that the variance of $\bar{y}$ is $1/\sum w_i$. Show that this reduces, as it must, to V/N when every $V_i = V$.

A warning—if variances estimated from different mean squares are used to calculate weights, the weighted mean $\bar{y}$ may be biased by the inaccuracy of estimation of the variances.

2.10. This example uses the rules for weighted analysis in Example 2.9 to analyse the non-orthogonal table of Example 2.5, taking advantage of the fact that the table has only two rows, i.e. male and female. The within-cells remainder mean square is 1.373. The average difference (male–female) for the British is

$$\frac{10.4 + 12.5}{2} - \frac{7.8 + 7.0 + 5.9 + 6.0}{4}$$

with variance $1.373(1/2 + 1/4)$, i.e. it is 4.775 with variance 1.0298 and therefore with weight $1/1.0298 = 0.9711$. Similarly the average difference (male–female) for Germans is 4.333 with weight 1.0925. There is no Australian comparison because there are no Australian women in the sample. The weighted average difference is therefore

$$(0.9711 \times 4.775 + 1.0925 \times 4.333)/(0.9711 + 1.0925) = 4.541$$

with variance $1/(0.9711 + 1.0925) = 0.4846$. This estimate of sex difference has excluded nationalities, i.e. it is 'sexes adj. nationalities'. Chapter 6 will show that t^2 must equal F when both ask the same question. Compare the value of t, $4.541/\sqrt{0.4846}$, with the corresponding variance ratio viz. 'sexes adj. nationalities'/remainder, arising from the analysis of variance of Example 2.5. How can you obtain a value of t corresponding to the variance ratio interactions/remainder?

3 Regression, correlation, and determination

Chapter 1 discussed the use of some function F_i—which might be a mean, a regression, or any other suitable function—to predict the value of y_i. In Chapter 2 the emphasis shifted, so that we considered means as entities in their own right. This chapter considers regressions in the same way.

The estimation of means may be considered as a special case of regression. Suppose we have a one-way table of values of y. For example, y might be the milk yield of a cow, and the 'treatments' might be different breeds, Friesian, Guernsey, Jersey, and Shorthorn. To estimate the average yield of each breed, we use the model

$$y = \text{the appropriate treatment mean} + \text{remainder.} \qquad (3.1)$$

We shall now create 'dummy variates', one for each breed. Usually a variate is something we measure, e.g. the live weight of a cow. But a dummy variate is something quite arbitrary; we ourselves decide what its values shall be. In this case we use it to code for a breed of cow. The assigned value of x_1 is 1 for all Friesian cows and 0 for cows of any other breed. The assigned value of x_2 is 1 for Guernseys and 0 for any other breed. Similarly x_3 and x_4 indicate Jerseys and Shorthorns. So if a cow is Friesian, it has $x_1 = 1$ and $x_2 = x_3 = x_4 = 0$; and so on. Then eqn (3.1) can be rewritten

$$y = b_1 x_1 + b_2 x_2 + b_3 x_3 + b_4 x_4 + \text{remainder.} \qquad (3.2)$$

Since $x_1 = 1$ and $x_2 = x_3 = x_4 = 0$ for Friesians, eqn (3.2) becomes $y = b_1 + \text{remainder}$ for Friesians, i.e. b_1 is the Friesian mean. Similarly b_2 is the Guernsey mean, and so on. But eqn (3.2) is the basic equation for multiple regression. If we fed the variates y and x_1, x_2, x_3, x_4 into a multiple regression program, we should get values of b_1, b_2, b_3, b_4 identical to the sample means for each breed. We should be doing the same analysis by a different method, and so we must get the same answer.

Dummy variates are mentioned for two reasons. First, they show that there is no intrinsic difference between the fitting of means and the fitting of regressions. It is true that in a regression the x-variates usually take a continuous range of values, whereas we fit means to discrete categories such as Friesians or Guernseys, so that the dummy x-variates in eqn (3.2) take discrete values 0 and 1; but that makes no difference to the regression analysis itself. Secondly, as we shall see, dummy variates are sometimes used in practice. Usually we calculate means directly, because they require much less computation than a multiple regression. But sometimes when dealing with a lot of non-orthogonal cross-treatments, it saves trouble to use dummy x-variates, assuming that you have an adequate multiple regression program.

3.1. Single regressions

The simplest regression equation, a straight line of slope b, is

$$y = bx + \text{remainder.} \tag{3.3}$$

This equation deliberately contains no intercept. If $x = 0$, y will on average be zero, and so the regression line goes through the origin. To estimate b we minimize the sum of squares $\sum(y - bx)^2$. Elementary calculus shows that b is estimated as $\sum xy / \sum x^2$. The values of $\sum xy$ and $\sum x^2$ are calculated from the original values of the xs and ys. This is the case mentioned in Chapter 1, where neither y nor x is corrected for the mean as a first step in the analysis. Assuming that the intercept really is zero, the sample mean $\bar{y}$ gives no extra prediction of a future value of y_i, once bx_i is known. In fact, the best predictor of y_i is simply bx_i (eqn 3.3). Consequently the analysis of variance of y includes no correction for the mean, even though $\bar{y}$ may be significantly different from zero. In such cases, the 'sum of squares for fitting the mean $\bar{y}$' is wholly absorbed by the regression on x.

Usually, however, we include an intercept a, to get the ordinary linear regression of y on a single x-variate:

$$y = a + bx + \text{remainder.} \tag{3.4}$$

Strictly speaking, since eqn (3.4) contains *two* coefficients a and b, it could be regarded as a regression on *two* x-variates, one of which is x itself (with coefficient b) and the other a dummy variate (coefficient a) which always take the value 1. But since in practice it is very simple to estimate the extra parameter a, we call eqn (3.4) a single regression. It can be shown[1] that to minimize $\sum(y - a - bx)^2$ is equivalent to minimizing $\sum[(y - \bar{y}) - b(x - \bar{x})]^2$. Therefore, b is estimated as

$$\sum(x - \bar{x})(y - \bar{y}) / \sum(x - \bar{x})^2,$$

so that the analysis is much the same as for eqn (3.3) except that both x and y are corrected for their respective means. In other words, b is calculated not from x and y as such, but from their deviations from the mean.

In practice, eqn (3.4) is almost always used in preference to eqn (3.3).

Suppose we are concerned with the regression of body weight on body length in adult sheep. It is true that a sheep of zero length would have zero weight, i.e. the regression must theoretically go through the origin. But we can't expect the relation between weight and length to be a straight line over the whole range of lengths from zero to adult size; whereas it may or may not be possible to describe the relation, over the range of adult sizes which concern us, by a straight line which does *not* go through the origin. We need not insist that a regression line must go through the origin unless the values of y and x actually extend down towards zero. Throughout this chapter we assume that the relation between y and x is, over the range of values in the sample, a straight line. Chapters 4 and 5 deal with problems of non-linearity.[2]

Regression is essentially a method for predicting y from x. For example, the weights of eye lenses are sometimes used to estimate the ages of wild animals, because the growth of the eye lens is not greatly affected by environmental disturbances. Obviously, the size of the eye lens depends on the age of the animal, rather than vice versa. So it might seem appropriate to regress lens weight on age. But we want to predict the animals' ages from their lens weights; for that purpose we regress age on lens weight.[3] In general the regression lines of y on x, and of x on y, are different. That need cause no difficulty: it means that the best formulae for predicting y from x, and x from y, are not the same.

3.1.1. Functional relationship

Regression is used to *predict y* from *x*. Most often, that is precisely what we want to do, even when it might seem at first sight that we really want to *describe* the biological relation between y and x. But sometimes we know in advance that there will be a linear 'functional relation' between two variates, and we want to estimate the parameters a and b of that relation. We cannot often be sure in biology that the relation will be truly linear, but, for example, we can expect enzyme activity to be directly proportional to enzyme concentration over a certain range of values. We know that there are *two* regression lines, of y on x (predicting y from x) and of x on y (predicting x from y), but there can be only *one* underlying functional relation. How can we find it?

A difficulty arises because of errors of measurement. Suppose the observed values y and x differ from the true biological values Y and X by unavoidable errors of measurement, e.g. we can't measure the true enzyme concentrations exactly. At one extreme, the errors of measurement are zero, in which case the values of $y(=Y)$ and $x(=X)$ must lie exactly on the straight line 'functional relation', and both regression lines will coincide with the functional relation. At the other extreme, errors of measurement will be large so that the regression of y on x gives poor predictions of y, yet the underlying relation between Y and X remains untouched.[4] The technical problems of estimating a functional relation are discussed in the next two paragraphs. They do not affect the normal use of regression to predict y from x, so you may wish to skip the next two paragraphs.

The true values of X and Y exactly obey the equation

$$Y = A + BX.$$

We want to find A and B, but we don't know the values of the Xs and Ys. The regression coefficient b of y on x is $\sum(y-\bar{y})(x-\bar{x})/\sum(x-\bar{x})^2$. Suppose that the sample size is N, and the variance of the errors of x is V. It can be shown that $\sum(y-\bar{y})(x-\bar{x})$ will on average equal $\sum(Y-\bar{Y})(X-\bar{X})$, but $\sum(x-\bar{x})^2$ will on average equal $\sum(X-\bar{X})^2 + (N-1)V$, and so the regression coefficient of y on x is

$$b = \frac{\sum(y-\bar{y})(x-\bar{x})}{\sum(x-\bar{x})^2} \cong \frac{\sum(Y-\bar{Y})(X-\bar{X})}{\sum(X-\bar{X})^2 + (N-1)V}.$$

But the unknown value B is $\sum(Y-\bar{Y})(X-\bar{X})/\sum(X-\bar{X})^2$. Unless $V=0$, the denominator of b, viz. $\sum(X-\bar{X})^2 + (N-1)V$, will exceed the denominator of B, viz. $\sum(X-\bar{X})^2$, and so the absolute size of b comes out smaller than B. Unless $V=0$, i.e. unless x is measured without error, the regression of y on x gives a diluted estimate of the slope of Y on X. So if x is known without error, the regression of y on x estimates the functional relation between Y and X—assuming that an exact linear relation exists; and if y is known without error, the regression of x on y does so. But if both x and y have errors of measurement, the line $Y = A + BX$ lies somewhere between the regression lines of y on x, and of x on y.

There are two practical remedies. If we can estimate V, we can correct the bias in b by subtracting $(N-1)V$ from the denominator $\sum(x-\bar{x})^2$; but if we do that, we are estimating $\sum(X-\bar{X})^2$ as a 'component of variance',[5] which is often rather risky (Chapters 4 and 10). The other way of estimating B is to split the sample into two halves, one containing all the large values of x and y and one containing all the small values: find the means $(\bar{x}, \bar{y})$ for each half: and calculate the slope of the line joining those two 'centres of gravity'. This method cannot be used on every sample. It requires that when the sample is split into large and small values of x, it shall simultaneously split into large and small values of y. When that is not possible—or very nearly so—the method should not be used. The whole difficulty, that b is a biased estimate of B, may also be avoided as follows. An experimenter decides what the values of X shall be, and arranges the experiment accordingly. For example, X might be a desired setting of an instrument, and x the setting actually achieved. The value of x will in general differ from X, but as long as positive and negative errors are equally likely, it is legitimate to analyse the results with X in place of x. The basic point here is that the experimenter himself prescribes—and therefore knows—what X should be.[6]

We now return to the predictive use of regression. Fifty years ago, statisticians thought in terms of correlations which treat x and y as equal partners, rather than regressions. Nowadays we usually use regressions, recognizing that most statistical problems are really problems of prediction. Since the x-variates are used only as 'something which will, we hope, predict y', they can be anything we choose. They can be a pair of old boots if it will predict y. If they include errors of measurement, it doesn't matter because any future values of x used to predict corresponding values of y will include similar errors of measurement.[4] The xs need not be distributed Normally, or in any other prescribed way. The regression analysis takes the values of x for granted and minimizes the sum of squares of the vertical remainders in Fig. 3.1, i.e. it is concerned to make the predictions of y as accurate as possible. It assumes that the remainder variance of y does not change as x changes; if it does, a weighted analysis would be more appropriate to allow for the fact that different ys have different accuracies.

We may use the rule given in Chapter 2 to calculate the standard error of the regression coefficient b. The value of b is $\sum(y-\bar{y})(x-\bar{x})/\sum(x-\bar{x})^2$. Since $\sum(x-\bar{x})=0$ by definition of $\bar{x}$, it follows

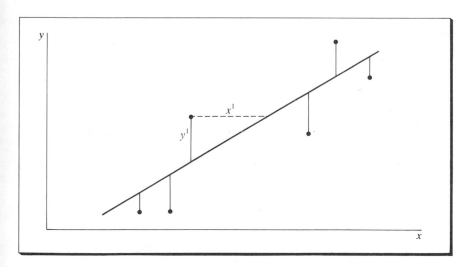

Fig. 3.1. The regression of y on x minimizes the sum of squares of the vertical remainders y'. The regression of x on y would minimize the sum of squares of the horizontal deviations x'. The two regressions are different unless the data points all lie exactly on a straight line, or are symmetrically arranged about one.

that $\bar{y}\sum(x-\bar{x})=0$. Therefore the value of b is equal to $\sum y(x-\bar{x})/\sum(x-\bar{x})^2$. So b is $a_1y_1+a_2y_2+\cdots$, where $a_i=(x_i-\bar{x})/\sum(x-\bar{x})^2$. It is because b is a *linear* function of the ys that regressions are 'robust' (Chapter 4). Since every y has the same remainder variance V, the variance of b is $a_1^2V+a_2^2V+\cdots$, which becomes $V/\sum(x-\bar{x})^2$. This is the usual expression for the variance of a regression coefficient. V is estimated by the remainder mean square of y, and the standard error of b is the square root of its variance, i.e. $\sqrt{[V/\sum(x-\bar{x})^2]}$.

3.1.2. Analysis of covariance

Sometimes we wish to compare the regression of y on x in different blocks of data. Blocks are sets of data which we keep distinct because we suspect that they might have different means $\bar{x}$ or $\bar{y}$, or different regression slopes b. Blocks might represent treatments of interest, e.g. male versus female, or they might represent unavoidable discontinuities—e.g. data collected at different places or in different years—which need to be eliminated from the analysis. We could of course analyse each block separately, but usually we analyse all the blocks together, to get a single analysis of variance with a single residual mean square.

We work on the usual plan described in Chapter 1. First we ask if there is any overall regression (within blocks), and then we fit a separate regression for each block, to see if the regression differs from block to block. The situation is complicated by the fact that a regression equation contains two coefficients a and b, which can be examined separately. But we are usually more interested in the slope b than in the intercept a. It is the slope that says how much change to expect in y, consequent on any given change in x. If two regression lines have different slopes, the regressions are intrinsically different. In that case the two lines must meet somewhere, but precisely where is usually of no great interest. There is then no point in asking whether the intercepts are the same, i.e. whether the two lines meet at $x=0$. If the slopes are different, we don't care too much about the intercepts; but if the slopes are the same, we shall want to know whether the lines are parallel (intercepts different) or identical (intercepts equal).

So we proceed as follows. We correct every value of x and y for its block mean, so that average differences between blocks are eliminated. Then we calculate the overall regression slope to see if x can predict y at all. Then we calculate a regression for each block separately, to see if the slope differs from block to block. If so, we need to use a different equation, both slope and intercept, for each block. But if the slope is approximately the same in every block, we adopt the overall value of b and then ask if the intercepts are different from block to block. It can happen that different slopes in different blocks cancel each other so that there is no overall regression, but such cases are very rare. All these questions may be examined by analysis of variance of y, or equivalently by looking at the values of a and b with their standard errors.

Now suppose that the different blocks have different values of $\bar{x}$ and $\bar{y}$, but the regression slope is the same in every block. Then we can, if desired, use the regression to correct the values of $\bar{y}$ for the differences in $\bar{x}$, i.e. compare the $\bar{y}$s at a standard value of $\bar{x}$, say x_0. Figure 3.2 shows the regression in one block. Starting from the observed values $(\bar{x}, \bar{y})$, the corrected value of $\bar{y}$ is found by moving along the regression line from $x = \bar{x}$ to $x = x_0$. Since the regression slope is the same in every block, the comparison between the corrected values of $\bar{y}$ will be the same, whatever standard value x_0 we may use. The method should not be used when the slopes are not the same. If we use $x_0 = 0$, the corrected $\bar{y}$ is simply the intercept. So if the regression lines are parallel in different blocks (intercepts different) there still remain differences between $\bar{y}$s after correction for the differences in $\bar{x}$; but if the regression lines in different blocks coincide (intercepts equal) the observed differences in $\bar{x}$ are sufficient to explain the observed differences in $\bar{y}$. In other words, the

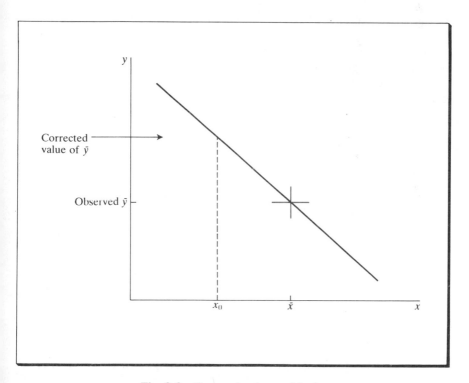

Fig. 3.2. Regression in one block.

between-blocks regression of $\bar{y}$s on $\bar{x}$s is the same as the within-blocks regression.

This method, called *analysis of covariance*, of correcting $\bar{y}$ to a common $\bar{x}$ can be very misleading when 'blocks' refer to experimental treatments, and those treatments affect x itself—for then we are using one treatment effect (on x) to correct another effect (on y) of the same treatment. The procedure is used only when it's certain that the experimental treatments can have no effect on x. For example, experimental treatments cannot affect the calendar age of an animal. If x is age, differences in x between treatments cannot be due to the treatments themselves, and so it is permissible to correct y for x.

Regression analysis, like any other statistical method, can be misused. Figure 3.3 shows three cases where mechanical use of the method could mislead. The commonest case is Fig. 3.3(a) where a linear regression is fitted to data which show important curvature. Predictions within the range of the original data are not as accurate as they might be, and extrapolation outside the range would be disastrous. Some regression programs are fitted with automatic tests for curvature, but the only sure way of detecting curvature is to *look at a plot of the data*. Figure 3.3(b) shows a sudden jump from 0 to 100 per cent. A regression of y on x would be highly significant, but it would not represent the true relation between x and y. In Fig. 3.3(c) there is clearly something unusual about the single outstanding point. It might even be erroneous, yet it largely determines the position of the regression line. It would be dangerous to use the regression in Fig. 3.3(c) to interpolate for intermediate values of x, since there is no evidence that the relation between x and y is truly

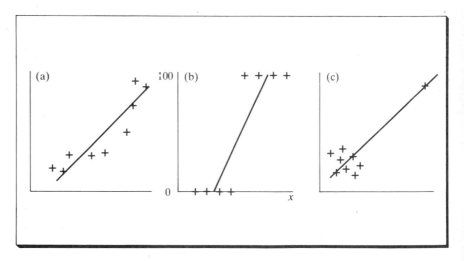

Fig. 3.3. The importance of plotting graphs.

linear. Unless there is some theory which specifies what the relation between y and x should look like, we can never safely extrapolate any regression, however well estimated, beyond the observed range of xs. Figure 3.3 shows how very important it is to plot things graphically. It is much easier to detect irregularities in a graph or histogram than in sets of written numbers.

You should make sure that you can do Examples 3.1–3.3 before going on to multiple regression.

3.2. Multiple regression

Multiple regression of y on several xs uses the equation

$$y = a + b_1 x_1 + b_2 x_2 + \cdots + \text{remainder} \tag{3.5}$$

Equation (3.5) is the same as eqn (3.2) except that it includes an intercept a. Once again the primary purpose is to predict y from the xs. If the xs are not correlated among themselves, the multiple regression is a straightforward combination of all the single regressions of y on each x, the coefficients b in eqn (3.5) are the same as the single coefficients in eqn (3.4), and the multiple regression accounts for a sum of squares of y equal to the total of the regression sums of squares of all the single regressions. That situation is analogous to the orthogonal two-way table of Chapter 2, where the rows sum of squares and the columns sum of squares add up to give the rows + columns sum of squares. In real life the xs are themselves correlated; it doesn't matter whether their correlations are significant or not. Then the multiple regression has to allow for the correlations between the xs. The situation is analogous to a non-orthogonal two-way table, where we have to adjust rows for columns and vice versa, and where the rows + columns sum of squares is not the total of the straight rows and columns sums of squares. The multiple regression coefficients b in eqn (3.5) are no longer the same as the single coefficients b in eqn (3.4), and the multiple regression accounts for a sum of squares of y which is not the total of the regression sums of squares of all the single regressions. An extreme case would be where two x-variates x_1 and x_2 were always identical (correlation = 1), in which case a multiple regression of y on x_1 and x_2 could give no better prediction of y than a single regression on either x_1 or x_2, and so the sum of squares would be no bigger than the sum of squares for the single regression. In any case, whatever the xs may be, the multiple regression of eqn (3.5) chooses the bs which give the best linear prediction of y from that particular set of xs.

We can usually make sure that a two-way table shall be orthogonal. A non-orthogonal two-way table is a rarity that is best avoided wherever possible. But in a multiple regression we usually cannot ensure that the xs shall be completely uncorrelated, and so in practice multiple regressions are nearly always complicated by correlations between the xs.

Suppose we wish to investigate the regression relations between many variates. There will be a vast number of possible regressions, using all possible combinations of one, two, three ... x-variates. The choice of which regressions to look at will be dictated by the biological questions to be asked. Now that computers have made it easy to churn out multiple regressions *ad nauseam*, it is very important to keep those biological questions firmly in mind. But suppose we have chosen a y-variate for biological reasons, and there are many possible x-variates which might be used to predict y. We may wish to find the smallest possible sub-set of the xs which can predict y as accurately as can the whole set of xs. In theory we might try all possible combinations (each x absent or present); but in practice there are too many possible combinations, if there are more than five or six xs. Sometimes there are biological considerations which recommend some xs above others, or some xs may be easier to measure. If not, two different sets of xs are equally acceptable if they predict y equally well.

There are no hard-and-fast rules for choosing trial combinations of xs; the job is a matter of judgement and commonsense. Two common methods are 'stripping down', where we start with the regression on all the xs together and step-by-step remove the 'smallest' x-variate, i.e. that x which contributes least to the regression; and 'building up', where we start with no xs and step-by-step add in the 'best' x-variate. As always, the remainder mean square tells us how much of the original variance of y remains unaccounted for by the regression. When stripping down, we know the size of the remainder mean square when all xs are included in the regression, and so we stop removing x-variates when the remainder mean square seriously increases. When building up, we stop adding xs when the remainder mean square no longer decreases seriously.

Neither method is foolproof. Unless the xs are all uncorrelated, there is no guarantee that either method will lead to the optimal combination of xs. The order in which xs are included when building up often differs from the order in which they are removed during stripping down—one process is not the reverse of the other. That is because unless the xs are uncorrelated, each x does not account for a unique amount of the variance of y. The importance of any x depends to some extent on which other xs have been included in the regression. When stripping down, it is safe to discard any x whose regression coefficient is less than its standard error, but only one such variate at a time, because, oddly enough, removal of one unimportant x-variate may make another one important.

The essential criterion used to judge the success of a regression is the reduction it achieves in the residual mean square; but to *test* that reduction we use the variance ratio of 'regression' mean square to residual mean square (Chapter 1). Now suppose we wish to test the significance of the *extra* prediction achieved by including an extra x-variate. For example, is the regression on x_1, x_3 and x_4 better than on x_1 and x_4 alone? We use the method of Chapter 1, taking the regression on x_1 and x_4 for granted. That is, we analyse the sum of squares of y remaining after the regression on x_1 and x_4 into two parts: the remainder sum of squares after regression on x_1, x_3 and x_4 and the 'sum of squares absorbed by the extra regression on x_3 after x_1 and x_4 have been fitted'. The variance ratio (F-) test then compares the two mean squares.

Similarly, if we want to test whether the regression on x_1, x_3 and x_4 is significantly better than regression on x_1, x_2 and x_4, we must compare the 'sum of squares absorbed by extra regression on x_3 after x_1 and x_4 have been fitted' with the 'sum of squares absorbed by extra regression on x_2 after x_1 and x_4 have been fitted'. Since those two sums of squares have only one degree of freedom each, their variance ratio will have to be very large before significance is established. But when examining regressions it is often not necessary to do significance tests; the behaviour of the residual mean square reveals the whole story.

Sometimes we wish to see if a multiple regression is the same within several blocks of data. The principle is the same as before. Does it pay to fit a separate regression for each block? Since the coefficients of a multiple regression are inter-related, it is rather meaningless to test the homogeneity of any one coefficient on its own. Rather, we test the homogeneity of the whole set of regression coefficients.

3.2.1. Determination

Some people use multiple regression not to predict y, but to identify those xs which determine y. This demands the greatest caution. If the data are observational rather than experimental, there can be *no* guarantee that any observed relation is cause-and-effect (Chapter 8). If it were known in advance that a given set of xs does determine y, multiple regression might be used to assess the relative importance of those xs, with three provisos. (1) As discussed above, errors of measurement of x tend to dilute the true size of the functional relation. (2) All the important xs must be included in the analysis; if one is missing and it is correlated with some of the other xs, those others will tend to fill the gap, giving an erroneous idea of their own direct importance. (3) The regression analysis is based on eqn (3.5), which assumes that the effects of the xs are linear and additive. If those assumptions are wrong, the analysis will wrongly estimate the relation between y and xs (Chapter 4).

The difficulties are even greater when we don't know in advance which

xs will determine y. In Chapter 2 we saw that if a two-way table is not orthogonal, it is impossible to sort out the effects of rows and columns completely. Precisely the same difficulty occurs in multiple regression unless the xs are uncorrelated. The coefficient b of any x depends on which other xs are included in the analysis. It is always possible that the inclusion of some x which has not actually been measured could profoundly alter the coefficients of the existing regression equation. 'It is probably unwise to try to assign relative determinations to correlated determining variables. Since in general determination is a complex thing, we do not lose much by failing to answer the question.' (Tukey 1954).

3.2.2. Discriminant analysis

Discriminant analysis is a special case of multiple regression. That fact is not generally known, because the theory of discriminant analysis was worked out before its relation to regression was recognized. We take samples from two different populations—which might for example be two species—and measure various variates x on the individuals in the samples. The choice of which xs to measure is entirely ours. We know which individuals belong to which population. We now wish to take future individuals and assign them to one or other population on the evidence of the xs. That is, we ask what is the linear combination of the xs that best discriminates between the two populations. Suppose we set up a dummy variate y which takes values 0 for all individuals of one population and 1 for the other. Any other pair of numbers would do as well as 0 and 1—it's the difference that is important. Then the discriminant function is the multiple regression of y on the xs, when all the data are grouped together in one block. In fact, we are using the xs to predict a value of 0 or 1 for each future individual. Here the distribution of y is clearly not Normal, but theory shows that if the xs are Normally distributed, the y-remainders will be Normal, so that the usual significance tests are valid. As mentioned above, usually in regressions the xs can have any distribution they like, provided that the y-residuals are Normal; in discriminant analysis the xs are taken to be Normal just to ensure that the y-residuals shall be Normal.

By actually using this dummy y-variate in a multiple-regression program, we avoid the need for a special program for discriminant analysis. The function which discriminates between *two* populations is indeed the best available linear function for the purpose. Such is no longer the case when we try to discriminate between three or more populations at once. There can be no guarantee that the function which best discriminates between two of the populations will also be the best to discriminate either of the two from a third population. Therefore, although there exist 'canonical analyses' to discriminate between more than two populations simultaneously, they are not always very satis-factory. Some textbooks reserve the term 'discriminant analysis' for the case of two populations only, while others use it for the more general case of two or more populations.

In discriminant analysis, we already know to which group (population) each individual in the sample belongs, and we use the discriminant function to assign subsequent individuals to those groups. But there also exist several methods of sorting the individuals in a sample into groups, purely by looking at the various measurements x made on those individuals. The groups are not defined in advance; instead the computer impartially sorts the sample into groups of similar individuals. A major difficulty is that you can get any grouping you like, by defining 'similar' accordingly. As an extreme case, different methods could sort male and female mice and elephants either into groups of males and females, or into mice and elephants. Methods of sorting must therefore be sub-jective, whether done by computer or not—but they can still be useful for some defined purpose. For example, taxonomic classification is to some extent subjective, but it is still very useful. Usually a computer cannot classify as well as a human expert, solely because the expert can't specify exactly how he does the job. If he could, the computer could be programmed to do the same thing.

3.2.3. Correlation

The idea of regression is closely connected with that of correlation. Correlation, like regression, considers the *linear* relation between two variates. For example, although there is a direct connection between x and x^2, they are not perfectly correlated because they don't lie on a straight line. As we have seen, regression is used to predict y from x, so that x and y are not on equal footing. The greater the correlation between x and y, the better the prediction will be. The correlation coefficient r measures the degree of linear co-relation between x and y regarded as equal partners, and the regression of y on x tells us how much to expect y to change (because of the co-relation) in response to any given change in x. Although the regression lines of y on x and of x on y are not identical, the degree of predictability is the same in each case. In fact, as explained in Section 6.2.2, either regression absorbs a fraction r^2 of the original sum of squares. A correlation coefficient always lies between -1 and $+1$, whereas a regression coefficient may take any value. Correlation coefficients used to be very important in statistical practice, but nowadays we usually think in terms of regression, which emphasizes the predictive aspect.

Nevertheless, correlation coefficients are far from dead. For example, they are important in the theory of selection. Suppose we wish to select individuals for some desirable trait y, but we can't measure y directly and so have to make do with a correlated variate x. Then unless the correlation between y and x is perfect ($r = +1$ or -1) there will be errors of selection, and the more stringent the selection, the greater the frequency of mistakes. For example, if $r = 0.6$ and we select the 5 per cent of individuals who have the largest values of x, 31 per cent (not 100 per cent) of those selected individuals would also have appeared in a selection for the top 5 per cent of ys: but if we select the top $\frac{1}{2}$ per cent of xs, only 15 per cent of the selected individuals would also have appeared in a selection for the top $\frac{1}{2}$ per cent of ys. So university examinations, which assess innate ability by criteria which are imperfectly correlated with that ability, cannot be altogether accurate! On the other hand, even if the correlation is very weak, mild selection for x can considerably increase the proportion of desirable ys.

3.2.4. *Principal components and factor analysis*

Sometimes we wish to describe the correlations between many variates, without using them to predict some y-variate by regression. If there are more than about six variates, we are faced with a mass of correlation coefficients. Thus ten variates have forty-five correlations between them. Unaided, we cannot grasp the pattern underlying so many individual correlations. Principal components analysis tries to summarize the set of correlations. These principal components have nothing to do with the components of variance mentioned in Note 5. The following paragraphs do not describe the analysis in detail—that analysis is done by specialized computer programs—but merely indicate what it is about.

A principal component is an additive combination $a_1 x_1 + a_2 x_2 + \cdots + a_n x_n$ of the n original x-variates, with coefficients $a_1, a_2, \ldots, a_n$ chosen so that $a_i a_j$ shall equal, as nearly as possible, the correlation r_{ij} between the ith and jth variates. If the agreement were perfect, the $n(n-1)/2$ correlations would be reduced to only n as. In practice the agreement is not perfect and the first principal component is chosen to make it as good as possible. Then a second principal component $b_1 x_1 + b_2 x_2 + \cdots + b_n x_n$ is chosen so that $b_i b_j$ shall approximate as closely as possible the remainders ($r_{ij} - a_i a_j$); and so on until a set of n principal components have been calculated which together will reproduce perfectly the observed set of correlations. But it is often found that the first two or three principal components suffice to summarize, near enough, the observed set of correlations, and therefore enable us to visualize how the variates hang together. For example, suppose we have a set of measurements on beef carcasses. One principal component which represents a measure of overall size, and a second component which represents 'degree of fatness', can reproduce rather well—but not perfectly—the whole pattern of correlation between several measurements. Instead of having to consider hundreds of individual correlations, a cattle breeder need think only of those two principal components.

It is tempting to suppose that those components measure some underlying biological process. Indeed, the very fact that it is possible to reduce an entire pattern of correlations to two components shows that the original variates must be interconnected biologically—a conclusion which will surprise nobody. But the components are additive linear combinations of the variates, and it is most unlikely that the biological processes which give rise to the variates act in additive linear fashion. Although we base the analysis on the additive linear model, the success of that analysis does not mean that the biology is itself linear and additive. So the components tell us nothing in detail about the underlying biology, but they do help us visualize how the variates are connected. Principal components analysis applies, not just to a set of correlations, but to any two-dimensional mathematical matrix. It can be, and is, used on covariances too.

Factor analysis is closely related to principal component analysis. A given number of 'factors' will summarize a pattern of correlations rather better than the same number of principal components. But the actual values of the factors vary according to the number of factors used in the analysis. If I choose two factors and you opt for three, your first two factors will differ from mine. No such difficulty arises with principal components. We can of course analyse a set of correlations for one factor, then for two and so on until we have enough factors to give a satisfactory approximation to the whole set of correlations. But many people are suspicious of an analysis whose result depends to some extent on an arbitrary cut-off imposed by the statistician himself. They prefer the more objective, if rather less efficient, principal components. Actually the two methods give very similar results. The major difficulties with either method are the lack of adequate significance tests (mathematical difficulties have so far restricted the tests available) and the very questionable assumptions of additivity and linearity (Chapter 4).

In summary, nearly all bread-and-butter statistical methods can if necessary be reduced to eqn (3.5). Therefore, nearly all bread-and-butter statistical methods are versions of regression analysis. In biology we take samples to find out how to predict unknown values of y in the population at large. Therefore, regression analysis, which predicts ys from xs, is generally far more important than functional relations, principal components or factor analysis. Prediction is one thing, but determination and causality are quite another.

Notes

1. $\sum(y-a-bx)^2$ is the same as $\sum(y-bx-a)^2$. The value of a that minimizes this sum of squares must be the mean of $(y-bx)$, just as the value of m that minimizes $\sum(y-m)^2$ is $\bar{y}$ (Example 1.2). So a is estimated as $\bar{y}-b\bar{x}$, and $\sum(y-a-bx)^2$ becomes $\sum[(y-\bar{y})-b(x-\bar{x})]^2$. To find the value of b which minimizes this sum of squares, we set the differential with regard to b equal to zero, which gives immediately

$$\sum(y-\bar{y})(x-\bar{x})-b\sum(x-\bar{x})^2=0.$$

Hence the usual expression for b.

2. It often happens that there is more than one y corresponding to each x. Thus in Example 3.2 there are several birds recorded for each distance flown. Suppose that, for any given value x_i of x, the mean of all the corresponding ys is $\bar{y}_i$. If the number of individual y_is is the same at each x_i, the regression of the y_is on x_i is identical to the regression of $\bar{y}_i$ on x_i. If the number of y_is is not the same at each x_i, each $\bar{y}_i$ needs to be weighted by n_i, the number of y_is of which $\bar{y}_i$ is the mean. This conclusion derives either from the fact that the variance of $\bar{y}_i$ is V/n_i (Chapter 2) or directly from the formulae for the coefficients a and b.

3. This point has caused much difficulty. Some biologists feel that it just doesn't make sense to regress age on lens weight. But from the statistical point of view, the biological fact that lens weight depends on age—not vice versa—is irrelevant. To predict age, we regress age on anything we can find that will give satisfactory prediction—in this case, lens weight. Lwin and Maritz (1982) use a much more complicated argument to arrive at the same conclusion.

4. If the errors of measurement of x are so large that they completely swamp the true values X, the data points will appear anywhere on the x-scale (whatever X may be) and the plot will show a random scatter of points. The regression line will be horizontal ($b=0$). This means that the best predictor of y is simply $\bar{y}$, whatever the value of x—or in other words, x gives no prediction of y. Yet the true functional relation between Y and X, if it exists, will be unaffected. In the common case where errors of measurement of x dilute the value of b, the regression of y on x still gives the best available predictions of y, provided that future values of x (used to predict y) are measured in the same way, and with the same sort of errors, as the xs in the sample used to estimate the regression. That is an implicit assumption of regression analysis.

5. Each value of $x=$ the true biological value $X+$ an error of measurement. In this case there is no doubt that the two terms strictly *add* together, because the error of measurement is defined as the difference between x and X. We now assume that the error is uncorrelated with X, i.e. that the same sizes of error attach to both large and small values of X. Then the variance of x will equal the true biological variance of $X+$ the error variance V. The mean square $\sum(x-\bar{x})^2/(N-1)$ which estimates the variance of x, accordingly splits up into $\sum(X-\bar{X})^2/(N-1)+V$. These two terms are called the biological and error components of the variance of x. Components of variance represent an optional

further dissection of the mean squares, after the analysis of variance (which derives the sums of squares) is completed. The analysis of variance is a purely arithmetic procedure which can be applied to any set of data whatsoever, according to the various classifications (rows, columns, regressions) of those data. It does not itself depend on any assumptions. The subsequent assumptions about Normality etc. are needed only if we subsequently want to test the mean squares. By contrast, the division of the mean squares into components of variance depends vitally on the assumption of zero correlation between additive contributions to y. So the initial analysis of variance is very reliable, but a subsequent dissection of the mean squares into components of variance can be misleading when the assumptions are wrong (Chapter 4).

6. In an article addressed to biologists, Ricker (1973) discusses at length the choice of regression lines in various contexts. He does not mention that many situations, at first sight involving functional relationships, are really predictive. For example, in measuring the effect of fertilizer on crop yield, our aim is essentially to predict what the response in yield will be, if we apply so much fertilizer. Ricker is willing to estimate a linear functional relationship (as opposed to an ordinary predictive regression) even when the data make it clear that no such underlying strict relationship can exist. Many people would consider such an exercise meaningless. Those two points acknowledged, Ricker's article is recommended.

Examples 3

3.1. A bacterial population was grown in a chemostat, starting from a few dozen cells. Successive samples gave the following counts.

Time from start (minutes)	20	40	60	90	120	180	240	300	360	420
Bacterial count (per cm^3 of medium)	47	62	73	103	220	537	1580	4500	9200	12 800

Regress (a) the bacterial count, (b) the logarithm of the count, on time. Plot both (a) and (b) against time and draw in the regression lines. Which regression represents the growth best? Does it represent it satisfactorily?

3.2. Humming birds were captured and weighed at various places on their migration route from Louisiana to Mexico.

In the following table both distances and body weights are subject to errors of measurement. Calculate the regressions of body weight on distance and vice versa, in the three categories. If any of the categories may be combined, combine them. What is the predicted weight of an adult humming bird after it has flown (a) 500 miles, (b) 2000 miles? What is the predicted distance flown by a juvenile bird weighing 2.0 g?

Category 1 Adult males		Category 2 Adult females		Category 3 Juveniles	
Distance flown (miles)	Body weight (g)	Distance flown (miles)	Body weight (g)	Distance flown (miles)	Body weight (g)
450	2.8	450	2.9	260	4.1
450	3.2	260	3.4	260	3.7
450	2.8	260	2.9	260	4.3
260	3.0	70	4.1	70	5.0
260	3.7	70	4.1	70	4.2
70	4.2	70	3.6	70	4.3
70	3.8	70	4.0	70	4.8
70	3.7	70	3.8	70	5.3
720	1.9	720	2.5	720	2.8
720	2.3	720	2.2	720	1.9
		720	2.0		

3.3. The regression of y on x is estimated from two separate blocks of data. The residual variance of y is the same in each block.
(a) What is the formula for the regression coefficient in block 1, and what is its variance?
(b) If we want to take the weighted average of the two estimates of b, what weights should we use?
(c) Therefore, what is the formula for the combined estimate of b?
(d) Is the combined estimate of b the same as the estimate obtained by lumping the two blocks of data together?
   Now consider the same question, but dealing with the block means of y instead of the regression b.

3.4. Repeat Example 2.2 by multiple regression on dummy variates. First group all the data together in one block. The simplest method would be that described in the text, i.e. to create one dummy variate for each lot of pigs (cf. breed of cows) and then use eqn (3.2) which contains no intercept. Multiple regression programs commonly use eqn (3.5) which includes an intercept a. There is then a technical difficulty. If we use a dummy variate for each lot of pigs, we shall be fitting *four* parameters a, b_1, b_2, b_3 to *three* lots of pigs. One of the parameters must be superfluous; it is a special case of linear dependence, discussed in Chapter 4. The difficulty is avoided by discarding any one of the three dummy variates. This makes the analysis lop-sided, but as we shall see, the answer comes out the same. This minor technical difficulty does not arise in an ordinary multiple regression, only when the xs are dummy variates. So use two dummy variates—one to indicate whether an animal is Large White or not, and one for Landrace. Include the data for the Large White × Landrace cross, for which the dummy xs will both take zero values, but omit the aberrant value 4.37. When you have found the regression equation of growth rate on the two dummy variates, insert the appropriate values of x into it to find the expected growth rates of all three lots of pigs, and compare the answers with the average growth rates. What would be wrong with using a single dummy variate which takes values $x = 0$ for Large White, $x = 1$ for Landrace and $x = 2$ for Large White $\times$ Landrace?

3.5. The numbers of storks' nests counted in Jingistan, and the numbers of babies born, were

Year	Nests	Babies
1910	418	12 312
1911	422	12 208
1912	440	12 857
1913	440	12 819
1914	442	13 204
1915	454	13 670
1916	471	13 538
1917	473	14 111
1918	488	14 364
1919	495	14 437
1920	522	14 503
1921	518	14 866
1922	525	15 376

Show that the number of babies may be predicted from the number of storks' nests, but that after the calendar year has been included as an x-variate, the number of nests gives no extra prediction of the number of babies.

3.6. Venetian cats steal fish and carry fleas:

Fish stolen in one year	Number of fleas on cat	Weight of cat (kg)
417	3	2.7
630	9	3.0
734	253	2.4
626	9	3.0
639	26	3.2
456	66	1.5
713	2	4.0
957	269	2.6

(a) Show that cats which carry most fleas tend to steal most fish. Do the data prove that cats steal a lot of fish because they have a lot of fleas, or that they have a lot of fleas because they steal a lot of fish?
(b) Show that the cat's weight cannot be predicted from the number of fish, or from the number of fleas, but
(c) the cat's weight can be predicted from the number of fish and the number of fleas in combination.

3.7. The following table shows subjective scores y of the suitability as holiday areas of a number of waterfront sites. At each site the angle of slope x (degrees) was measured at three successive distances from the water.

Score y	x_1	x_2	x_3
3	30	10	0
2	25	25	0
1.5	15	15	15
3	25	15	15
3	30	30	10
4	50	35	0
2	20	0	0
5	30	30	30
4	35	35	35
2	20	10	20
5	40	40	40
3	10	10	10

Construct new variates $x_1 + x_3$, $x_1 - x_3$, $x_1 + x_2 + x_3$ and $(x_1 + x_2 + x_3)/3$.

(a) Regress y on all combinations of x_1, x_2 and x_3 to see which is best.

(b) Show that the orders of preference of x_1, x_2 and x_3 when 'building up' and 'stripping down' are different.

(c) Show that regression on $x_1 + x_3$ and $x_1 - x_3$ together is equivalent to regression on x_1 and x_3 together. Show from the regression equation that such must always be the case.

(d) Show that in this case, $x_1 - x_3$ gives no extra prediction of y after $x_1 + x_3$ has been included in the regression.

(e) Compare the single regressions of y on $x_1 + x_2 + x_3$ and on $(x_1 + x_2 + x_3)/3$, and show that either gives as good prediction, in this case, as x_1, x_2 and x_3 together.

3.8. Repeat Example 2.5 by multiple regression on dummy variates. Use one dummy variate to indicate Australian or not, one to indicate British or not, and one to indicate male or not. Proceed as follows.

(a) Sort the data into cells and do a one-way between-cells analysis of variance, to get a 'between-cells' sum of squares and a remainder, or within-cells, sum of squares. The rest of the analysis will dissect the 'between-cells' sum of squares according to the row and column classifications.

  Lump all the data into one block. Then

(b) Perform multiple regression on the first two dummies, which absorbs 'sum of squares for nationalities, ignoring sex'.

(c) Perform regression on the third dummy on its own, which absorbs 'sum of squares for sex, ignoring nationalities'.

(d) Perform multiple regression on all three dummies, which absorbs 'sum of squares for nationalities and sex'.

  You now have all the ingredients for the analysis of variance.

(e) Subtract 'sum of squares for nationalities and sex' from 'between-cells sum of squares' to get 'interactions' sum of squares.

(f) Subtract 'sum of squares for nationalities, ignoring sex' from 'sum of squares for nationalities and sex' to get 'extra sum of squares for sex, given nationalities' i.e. 'sex adj. nationalities' (Fig. 2.2).

(g) Subtract 'sum of squares for sex, ignoring nationalities' from 'sum of squares for nationalities and sex' to get 'sum of squares for nationalities adj. sex'. Compare with the previous two-way analysis and with the least-squares theory of Chapter 1.

4 Additivity and linearity

Before we can do any statistical analysis, we must (consciously or unconsciously) have a model. A model is an equation of the type

$$y_i = F_i + \text{remainder}$$

on which the analysis is based. For example, in Chapter 2 we had

$$y_{ij} = m + a_i + b_j + \text{interaction} + \text{remainder} \tag{4.1}$$

for a two-way table, and in Chapter 3

$$y_i = a + bx_i + \text{remainder} \tag{4.2}$$

for a single regression. The method of analysis also depends on the distribution of the residuals. Until now we have used the method of least squares. For example, in eqn (4.2) we minimize the sum of squares $\sum(y_i - a - bx_i)^2/V_i$, where V_i is the variance of the residual of y_i. Before we can do the analysis we must know, or postulate, V_i. Very often we can assume that it is the same for every y_i, in which case we need only minimize the unweighted sum of squares $\sum(y_i - a - bx_i)^2$. If we suppose that V_i varies, we must minimize the weighted sum of squares $\sum w_i(y_i - a - bx_i)^2$, taking w_i to be $1/V_i$.

In Chapter 7 we shall see that the method of least squares is a particular case of 'maximum likelihood'. Sometimes it is necessary to abandon least squares and use likelihood directly. But in any case, once the model has been decided on and once the distribution of residuals has been determined, there can be only one correct method of analysis to answer any given biological question. True, the same analysis may be calculated in different ways, but the results will be the same. Given the model and the distribution of remainders, there is no room for 'we might try this method of analysis, or perhaps that one instead'. So if two statisticians hand out contradictory advice, it means that (a) they have not understood the biological situation or the questions to be asked, (b) they are really saying the same thing in different ways, or (c) they think that different models or different remainder distributions are appropriate. The method of analysis (even if non-parametric) inevitably depends on assumptions about the structure of the data, and so we need to make sure that those assumptions are correct or near enough to validate the analysis.

All the common statistical methods depend on additive models. Equation (4.2) contains two addition signs, and eqn (4.1) has four. There is usually no reason why the underlying biological processes should be additive, but as we shall see, additive models demand relatively simple calculations and theory. The everyday use of orthodox additive methods was laborious before the invention of desk calculators. Now that we have computers, orthodox analyses can be done very easily, and we could certainly try to develop more penetrating non-additive methods. But at present, computers are used mainly to proliferate additive analyses. That is due in part to conservatism, but very largely it is because additive methods are quite satisfactory for most purposes. Moreover it is very difficult to derive appropriate significance tests for non-additive models.[1]

Since the analysis depends on an additive model, we must be reasonably confident that the data do at least approximately conform to that model. If not, it may be advisable to transform the scale of measurement until they do (Chapter 5). It is usually a matter of experience and judgement to decide whether a set of data conforms well enough to an additive model, but sometimes the question can be tested statistically. For example, in a multiple regression of y on x_1 and x_2, we can construct an additional variate $x_3 = x_1 \times x_2$. If the extra regression on x_3 is significant, it means that the data do not completely satisfy the additive regression model

$$y = a + b_1 x_1 + b_2 x_2. \qquad (4.3)$$

Tukey (1949) developed the corresponding test for orthogonal two-way tables, and it appears in some textbooks. It is true that these tests only look for one particular pattern of non-additivity, but they can usually indicate the existence of serious departures from additivity, even if the actual pattern is different.

The question of additivity is still more critical in multiple regression (including discriminant functions) than in two-way tables. When analysing tables we can often transform to additivity or, equivalently, use an appropriate non-additive model—and in any case, the interactions measure the degree of non-additivity. In multiple regression the underlying eqn (4.3) assumes, purely for ease of calculation, that the contributions of the various xs add together. Usually there is no good reason why they should. But if we use the additive model in a seriously non-additive situation, the result will merely be some loss of prediction. So the non-additivity does not invalidate the analysis, but merely weakens it. The assumption of additivity is even more basic in principal component analysis—hence the warning in Chapter 3 that while the components may summarize a set of correlations, they should not be accepted as biological entities in their own right.

We shall see in Chapter 6 that significance tests assume the remainders to be Normally distributed, and in Chapter 7 that the method of least squares is most easily justified by the same assumption. There is a very close connection between additivity and Normality. The Central Limit Theorem tells us that if a quantity y is the additive sum of many independent contributions, then (under very weak conditions which are nearly always fulfilled) y will be Normally distributed. That theorem does not apply if instead y is (say) the multiplicative product of many independent contributions. Conversely, if y_1 and y_2 are Normally distributed, then $y_1 + y_2$ and $y_1 - y_2$ are also Normally distributed, but other non-additive combinations of y_1 and y_2 are not.

So in practice we work as follows. We make sure that it is reasonable to suppose that, on the chosen scale of measurement, the effects of treatments will be roughly additive. If not, we transform the scale of measurement. Then we use the argument developed in Chapter 5 that if the treatment effects are additive, the relatively small remainders are likely to be Normally distributed. (It does not follow that the remainder variance V_i will necessarily be homogeneous.) Certainly there are tests to see if the remainders really are Normal, but those tests are not often applied, first because they require a large body of data to distinguish any moderate departure from Normality, and then because a large body of data will rarely conform exactly to the Normal distribution! As we shall see in the next paragraph, that doesn't matter. In practice additive models give satisfactory results, provided that we do not take them too literally. The danger is to assume that because an additive model fits the data, the underlying biological processes must themselves act additively. Chapter 10 illustrates the unfortunate consequences of such an assumption.

We may well ask 'Is it not dangerous to use methods based on the assumption of Normally distributed remainders, without checking that assumption carefully in every case?' We rely heavily on the fact that many statistical methods are 'robust'—they do not go seriously wrong in the face of reasonable departures from Normality. Means, regressions and t-tests are reliable, provided that the distribution of y-remainders is unimodal (i.e. has only one peak or hump) and is not badly skew. But variances are much more sensitive. Usually that doesn't matter because we use variances merely to calculate standard errors—that is, to assess the accuracy of what concerns us most, namely the means or regressions. But sometimes the variances are analysed for their own sake. It is then easy, for several reasons, to get quite misleading results.

Suppose we have a sum of squares $\sum y^2$ with n degrees of freedom. The ys may be original observations, but usually they are residuals, in which case the number of degrees of freedom is less than the number of ys in the sum of squares. Suppose that the ys are Normally distributed with a true (population) mean of zero and variance V. Then $\sum y^2 / V$ has a χ^2 distribution with n degrees of freedom. That is, if we take many such sums of squares $\sum y^2$, the various values of $\sum y^2 / V$ will follow the χ^2

distribution, because that is how the χ^2 distribution is defined. When we use χ^2 in practice it is usually to analyse data which are whole-number counts. That is an approximate use of χ^2. The calculated values of χ^2 conform only approximately to the true χ^2 distribution, because the residuals used to calculate χ^2 are only approximately Normal. So we need to obey the rules which ensure that the approximation is a good one (Chapter 6). But χ^2 is basically the distribution of sums of squares of Normally distributed ys whose mean is zero. So if we take numerous sets of data and work out $\sum y^2$ for each set, the χ^2 distribution will tell us how much variation in $\sum y^2/V$ to expect. We can then deduce the variation in $\sum y^2$ itself.

The average, or expected value of χ^2 is always n, its number of degrees of freedom. For if $\sum y^2$ is a sum of squares with n degrees of freedom, the mean square $\sum y^2/n$ is an estimate of the population variance V. So $\sum y^2/n$ on average equals V, and therefore χ^2, i.e. $\sum y^2/V$, on average equals n. If we look at a table of χ^2, we see that the value of χ^2 becomes equal to the number of degrees of freedom somewhere near the middle of the probability distribution. Suppose for example that $n = 30$. The average value of χ^2 will be 30, but the table gives 18.493 for the 95 per cent probability point and 43.773 for the 5 per cent point. These two values are not symmetrical about the mean 30 because the χ^2 distribution is skew. So the 95 per cent and 5 per cent points of $\chi^2/30$ are 18.493/30 and 43.773/30, i.e. 0.62 and 1.46. That is, of a series of mean squares (each with 30 degrees of freedom) intended to estimate a variance V, 5 per cent will come out less than $0.62\,V$ and another 5 per cent will be more than $1.46\,V$, and so the estimate of V will often be very inaccurate. If we found that the estimate of a mean had a standard error of 20 per cent, we should think that estimate very inaccurate or even worthless in many cases, yet that is the kind of accuracy that must be expected from mean squares. If we are interested in the size of the variance for its own sake, a mean square with as many as 30 degrees of freedom may easily give a very inaccurate estimate. (Usually we calculate a mean square merely to assess the accuracy of a mean or regression coefficient. The published tables of t then make automatic allowance for the inaccuracy of the mean square as an estimate of V.) With fewer degrees of freedom, estimates of variance are even worse.

That is the first intrinsic difficulty in analysing variances *per se*. The second is that estimates of variance are not robust, i.e. departures from Normality tend to make them even more inaccurate. For instance, one aberrant value in a sample will not change the mean too much, but will enormously distort the mean square (Example 2.2).

Thirdly, when analysing variances in their own right, we may want to estimate not the variances themselves, but the 'components of variance' (Note 5, Chapter 3). These components of variance are not used in everyday statistical practice, and it's doubtful if they are much use anyway. Chapter 10 discusses their use in genetics. After an analysis of variance has split the total sum of squares into two or more sums of squares, the corresponding mean squares may (if desired) themselves be interpreted as the sums of components which are supposed to represent various kinds of biological variability. But there are serious difficulties. First, the components are rather different for a 'fixed-effects' model and for a 'random-effects' model. We have already met that distinction in Chapter 2. A fixed-effects model is concerned only with the particular treatments being analysed, whereas a random-effects model regards the treatments as a random sample from some population of possible treatments. It is often very difficult to decide whether a fixed- or random-effects model is most appropriate to a given case, and it is usually hard to swallow the assertion that the treatments which have actually been applied can be regarded as a purely random selection from some hypothetical population of treatments. Fortunately the distinction between fixed and random effects makes little difference to the answer in practice. Secondly, the components are found by subtracting one mean square from another. But if one mean square may estimate the true value inaccurately, the difference between two mean squares will be even less reliable. Indeed, components of variance sometimes come out negative, which in theory should be impossible. Thirdly, estimates of components of variance depend critically on the assumption of additivity of the treatment effects. Even the experts can go badly wrong when trying to interpret variances *per se*. That will not, perhaps, surprise people who have much experience of experts. None of these difficulties arise when variances are used in the usual way to assess the accuracy of means or regressions.

In Chapter 2 we observed that to describe a whole distribution, both the mean and the variance—or alternatively, the standard deviation— must be quoted. Occasionally, then, we want to analyse the variance of y in much the same way as we analyse its mean, but without getting involved in the complexities of 'components of variance'. In plant breeding, for example, reliability of yield is as important as average size of yield. A variety may be chosen because its performance does not vary much from year to year or from place to place. We may use a mean square to estimate the variability of a variety, just as we use the mean to estimate its yield. Since the distribution of χ^2 is very skew, it is best to apply a transform and analyse log(mean square). For example, Bartlett's test examines the homogeneity of a set of mean squares by comparing their logarithms. We still cannot escape from the fact that variances are not robust; Bartlett's test is very sensitive to departures from Normality. Instead of log(mean square), the cube root transformation $3\sqrt{}$(mean square) is equally satisfactory. It is also possible to regress log(mean square of y) on some concomitant variate x, to see if x can predict the variability of y. As a special case, x may be the mean $\bar{y}$ itself.[2] A further difficulty when analysing variances as such is that the variance commonly increases automatically as the mean increases, so that the difference between two variances may merely reflect the corresponding difference between the means.[2]

Additive models such as eqn (4.2) quite naturally lead to the use of regression methods, since eqn (4.2) *is* the basic regression equation. It assumes that the relation between y and x is a straight line, but the data may not agree. Linearity is a special type of additivity, since it says that each additional unit of x adds the same amount to y; but additivity and linearity are not synonymous. It is perfectly possible, in (say) a fertilizer trial, to have row and column effects which are themselves linear responses to increasing doses of nitrogen and potassium respectively, but which nevertheless interact, so that the effects of the two fertilizers are not perfectly additive. Just as, earlier in this chapter, it was possible to test for non-additivity by an extra regression on $x_1 \times x_2$, so it is possible to test for non-linearity by an extra regression on x^2. It does not follow that, if the test detects some curvature, that curvature is necessarily quadratic. The only sure way to see what kind of curvature is involved is to plot y against x graphically. If the curvature is slight, it may be good enough to ignore it and say 'the linear regression gives predictions which, although not perfect, are good enough for present purposes'. Otherwise there are several ways of dealing with non-linearity. Examination of the plot of y against x often suggests that some transformation of y or x will convert the relation to a straight line. A transformation of y will not only straighten the relation between y and x, but will alter the structure of the y-remainders, which is sometimes desirable (Chapter 5). On the other hand, a transformation of x should leave the y-residuals unaltered in structure (Example 5.3).

It is best to convert the situation to a straight line if possible, partly because we can then use linear regressions, and partly because it is easier to think in terms of linearity. But another way of dealing with curvature is to use a polynomial regression. This is the multiple regression of y on x, x^2, x^3, ... to as many terms as may be necessary to fit the observed relation between y and x. Two practical points arise.

First, the values of x, x^2, x^3 etc. are themselves highly correlated. To avoid numerical errors arising in the computation, it is best to use $(x - \bar{x})$, $(x - \bar{x})^2$, $(x - \bar{x})^3$, ... instead. These two methods of computation are equivalent, and give the same predictions of y. If the polynomial goes beyond $(x - \bar{x})^5$, even more sophisticated manipulation may be advisable. When the values of x are evenly spaced along the x-axis, the difficulty may be avoided by using orthogonal polynomials (Fisher and Yates 1963). These orthogonal polynomials were originally introduced to simplify the calculation; but even now when we have computers to do the job, they can still improve the accuracy.

Secondly, it sometimes happens that in a polynomial, some power of x less than the highest contributes insignificantly to the regression, so that it could apparently be omitted without loss. For reasons of mathematical consistency, that is not a good idea; the consequent adjustment to the other terms in the regression might distort the prediction of y at some point in the range of x. It is of course very dangerous to extrapolate any empirical prediction formula outside the range of values of x from which it was estimated. That is especially true of polynomials, which can bend about disconcertingly as soon as they move outside the original range of x. Such behaviour can wreck a simulation model which incorporates an empirical polynomial, and which unwittingly goes outside the original range.

Polynomials are useful but inelegant. Many terms may have to be included before a polynomial can predict the values of y as well as can some simple algebraic function of x, such as for example $x \log(x)$. Many people prefer to use such algebraic functions wherever possible, reserving polynomials as a last resort—a sledgehammer which can crack any nut. The choice of appropriate functions of x is discussed in Chapter 13.

But sometimes, no simple transform to linearity is possible. For example, we might decide to fit the relation

$$y = a - be^{-kx} + \text{remainder}$$

to a set of data. That relation cannot be transformed to a linear regression from which the values of a, b and k could be estimated. We must resort to a third method of dealing with curvature, viz. non-linear regression. The principle is the same as in ordinary regression. We find the values of a, b and k which minimize the remainder sum of squares $\sum(y - a + be^{-kx})^2$. But the calculation can no longer be done in one stage. It is necessary to find a, b and k in several steps, by successive iterations starting with arbitrary trial values. There are several possible methods of calculation. No one method is best in all cases, and indeed if the wrong method is chosen in any given case, the calculation can go on *ad infinitum*, without ever converging on the right values of a, b and k. Special computer programs exist to do the job, but it is often good enough to use an ordinary regression program. In this case, we might choose a series of trial values of k and find the values of a and b corresponding to each value of k by regression of y on e^{-kx}. We then choose the value of k which gives the best prediction of y (Example 13.2).

Suppose we wish to estimate a regression, and can ourselves choose the values of x before we measure the corresponding ys. The most accurate estimate of b is obtained by making half the xs as large as possible, and the other half as small as possible. There are two ways of seeing this. First, the regression line is anchored better by extreme values than by values near the centre of the graph. Second, the variance of b is estimated by $s^2/\sum(x_i - \bar{x})^2$ (Chapter 3). To make that variance small, $\sum(x_i - \bar{x})^2$ should be large, i.e. the values of $(x_i - \bar{x})^2$ must all be as large as possible. If it were absolutely certain that the regression must be linear, it would be best to choose the xs in that way. But then the curvature of the regression could not be examined, so it is usually best to include some intermediate values of x. If a gap is left in the range of values of x, predictions of y in the middle of that gap may be inaccurate.

If we are interested in two variates x_1 and x_2, we may find that the difference $x_1 - x_2$ is even more interesting. For example, the male and female flowers of a corn plant mature at different times, so that the plant cannot fertilize itself. If x_1 and x_2 are the times of male and female flowering, the difference $x_1 - x_2$ is genetically important. We wish to predict some variate y from the xs. It might seem reasonable to write $x_3 = x_1 - x_2$ and regress y on x_1, x_2 and x_3—but it won't work. The three x-variates are linearly dependent, i.e. any one of them may be written as a linear combination of the others. Consequently, any one of them may be omitted without loss. It is easy to see why. If

$$y = a + b_1 x_1 + b_2 x_2 + b_3 x_3 + \text{remainder}$$

then
$$y = a + b_1 x_1 + b_2 x_2 + b_3(x_1 - x_2) + \text{remainder},$$

i.e.
$$y = a + (b_1 + b_3)x_1 + (b_2 - b_3)x_2 + \text{remainder}.$$

The regression on x_1, x_2 and x_3 is equivalent to regression on x_1 and x_2 only. Any x-variate which is a linear combination of previous x-variates is superfluous. A multiple regression program should recognize and comment on such a situation. But sometimes a set of x-variates is almost, but not quite, linearly dependent. For example, that may happen if we regress y on x, x^2, x^3 without first subtracting the mean $\bar{x}$ from the values of x (see above). The analysis then becomes very unstable; small errors in the values of x may seriously affect the regression coefficients, although the prediction of y will be unaffected. The standard errors of some of the regression coefficients then become very large, and the situation may be recognized in that way.[3] Because of the way it does its arithmetic, a computer may fail to recognize that a set of x-variates is linearly dependent, in which case the results will show the same warning signals (very large standard errors) as if the xs were *almost* linearly dependent. This question of almost-linear-dependence is a natural hazard in multiple regressions, so it is important to do Example 4.1 to get the point.

Besides additivity, the other basic assumption in everyday statistics is that the remainders $(y_i - F_i)$ are uncorrelated with each other, i.e. there is no tendency for one of them to be large, just because another is large. This assumption is implicit in our use of the sum of squares $\sum(y_i - F_i)^2$, which would otherwise need to include cross-terms $(y_i - F_i)(y_j - F_j)$. The assumption is usually quite reasonable in biology, where measurements are taken on different biological units. For example, if we are weighing cats, there is no reason why one cat should be heavier than average, just because the previous cat happened to be heavy too. (If both cats are heavy because they have received the same experimental treatment, the average effect of that treatment will be represented in F_i, i.e. in the treatment mean, and not in the residuals.)

The situation changes when we take repeated measurements on the same biological unit. If the cat is too heavy today, it may well remain that way tomorrow, and so the successive residuals will be correlated. The difficulty does not apply to a before-and-after type of experiment, where each unit furnishes a single observation, namely the difference between start and finish. It does apply to a series of measurements taken on the same individual. The analysis of such 'time series' is a flourishing branch of statistics in its own right, but its main fields of application are dismal subjects like economics, and only rarely biology.

Another case where the assumption of independent residuals breaks down is agricultural field trials. No area of land is perfectly homogeneous, and if there are local patches of relatively fertile and infertile soil, the residuals of adjacent plots will be correlated when those plots happen to be located on the same patch. It is then possible to improve on the classical analysis, which assumes uncorrelated residuals, by allowing for the correlation between the residuals of neighbouring plots. Analyses of this type can only be done by computer, using special programs. There is currently a good deal of interest in such analyses (e.g. Wilkinson *et al.* 1983) but at present they offer little improvement over the classical solution to the problem, viz. the use of 'randomized incomplete block' designs which group the field plots into small blocks with treatments arranged at random within each block, and which eliminate the effects of 'patchiness' almost as well as do the special analyses.

All the same, we need to watch out for situations where residuals might indeed be correlated. The *reductio ad absurdum* would involve feeding one datum twice into the same analysis. It might appear that we could thereby increase the sample size at no extra cost, but of course there would be no gain in information; the two residuals would be identical and completely correlated. Yet the effect on the analysis, if the mistake were not detected, would be a reduction in the remainder mean square and an apparent increase in the accuracy of the means or regressions. This cautionary tale shows that although the assumption of uncorrelated residuals usually gives no trouble, it needs to be kept in mind.

Notes

1. Note 3 of Chapter 1 states that if we divide a sum of squares by its degrees of freedom, we get a mean square which estimates the population variance V. It can be shown theoretically that this is always true if we use an additive model, but need not be true if the model is not additive. Therefore, the standard variance-ratio test is not valid for analyses of variance based on non-additive models.

2. *Coefficient of variation.* As mentioned in Chapter 5, the variance of x often increases as the mean increases. If m_1 exceeds m_2, nobody will be surprised if the corresponding V_1 also exceeds V_2. Can we compare the intrinsic variability of two populations? Suppose the two means are sharply different, so that the ranges of x in the two populations do not overlap. Then it is possible to find a transformation of the scale of x to make V_1/V_2 take any value we like. So the comparison of the variances depends on the scale of measurement that is employed (Chapter 5). But if the two means are nearly equal, so that one population overlaps the other at both ends, the first population will be more variable than the second on any scale of measurement. We cannot say that one population is intrinsically more variable than another, unless the ranges of x overlap, or unless a particular scale of measurement is specified.

The 'coefficient of variation', equal to the standard deviation divided by $\bar{x}$, is a dimensionless measure of variability—that is, if we change the units of measurement from inches to centimetres, the coefficient of variation remains unaltered. We cannot use the coefficient of variation to compare the intrinsic variabilities of two populations with sharply different means unless we are sure that as m changes, V will remain proportional to m^2 (which it must if the coefficient of variation is not to change as m changes). In practice it is found that in a set of similar samples with different means, s^2 is proportional to $\bar{x}^b$, where b is some constant. This is the celebrated 'Taylor's power law' which Southwood (1978) discusses at some length, with ecological applications. Usually b does not equal 2, so that the coefficient of variation will change as m changes—in other words, the coefficient of variation can *not* be used to compare the variabilities of dissimilar populations, because it does not correct for the differences in $\bar{x}$ in an appropriate way. But $\log(s^2)$ may be predicted by linear regression on $\log(\bar{x})$, and then analysis of covariance (Chapter 3) may be used to correct differences in s^2 for differences in $\bar{x}$. This method evidently depends on having enough samples to estimate b accurately.

3. In the extreme case, suppose we regress on two identical variates x_1 and x_2, both always equal to a variate x. The single regression $y = a + bx$ is identical to

$$y = a + (b - \lambda)x_1 + \lambda x_2$$

for any value of λ whatever. The multiple regression coefficients $(b - \lambda)$ and λ are both arbitrary, and therefore have infinite standard errors; but their sum b is not and so the multiple regression equation still gives the same predictions of y, whatever the value of λ may chance to be. If a set of data of this type were fed into a multiple regression program, the program ought to recognize the linear dependence of the xs, but sometimes it will produce a multiple regression with an arbitrary value of λ and with very large standard errors for the regression coefficients.

Examples 4

4.1. The daily growth (increase in stem length) of a corn plant was measured on eight consecutive days, together with sunshine and rainfall. (Yes, maize plants really can grow at this rate!)

Growth (cm)	Hours of sunshine	Rain (in)
2.1	2.4	0.66
2.4	6.5	0.14
3.2	7.6	0.00
2.9	7.5	0.02
3.0	7.1	0.07
2.8	4.4	0.40
2.6	5.9	0.22
2.6	7.0	0.08

Examine the regression of growth on sunshine and rain. The analysis assumes that carry-over effects from day to day can be ignored. In this case the sunshine and rain records are highly correlated. They are nearly, but not quite, linearly dependent. Both give equally good prediction of growth, and the multiple regression of growth on sunshine and rain is erratic. Growth rate is actually *determined* neither by sunshine nor by rain, but by a third correlated variable, temperature.

5   Transformations

There is usually no reason why we should insist on analysing data on their original scale of measurement. If we are dealing with (say) numbers of animals, we can think just as well in terms of $\log(N)$ as of N itself. It is true that for some strictly practical things, e.g. pest numbers, we are directly concerned with numbers and not with the square roots of numbers. But even then it may be technically best to do the statistical analysis on a transformed scale, and finally change the answer back to the original scale.

5.1. Why use transformations?

There are three reasons for using transformations:

(1) to make the remainder variances uniform;

(2) to make the distribution of errors Normal;

(3) to make the effects of treatments additive.

1. Suppose we wish to compare the means of several blocks of data in a one-way analysis of variance. The usual analysis assumes that the remainder variance of single observations is the same in every block. That does not, of course, imply that the block *means* must all have the same accuracy, unless the number of observations per block is the same in every block. Fortunately, the method of analysis is insensitive—or robust—to reasonable departures from that assumption. And since the remainder variance within any block is usually correlated with the block mean itself (Note 2, Chapter 4) we need not worry too much, provided that the block means all have the same order of magnitude, and provided that the data have all been collected (measured) by the same method in every block. But if one block has mean 10 and another has mean 100— i.e. there is an order of magnitude difference between the blocks—it is unlikely that the remainder variances will be as great in the first block as in the second. This, in theory at any rate, invalidates the usual analysis, although the disparity has to be very great before it makes any real difference to the biological answer obtained. One way of rectifying the trouble is to transform the scale of observation. Thus, for a Poisson distribution, the variance of n equals the average value of n, and hence the variance of $\sqrt{n}$ is approximately independent of the level of size of n. Again, the Binomial distribution with proportion p has variance

$p(1-p)$, but the angular transformation gives almost uniform variance for all values of p between 0 and 1.[1]

2. If the distribution of remainders is not Normal, the analysis itself will be unaffected (unless the departure from Normality is so enormous that the analysis wastes most of the information contained in the sample—Chapter 7) but significance tests based on the Normal distribution will be upset. In the same way that estimates of means are robust (Chapter 4), so the significance tests of differences between those means are not greatly affected unless the departure from Normality is extreme: but tests on variances, e.g. variance ratio and Bartlett's test, are much more sensitive. Transformations may be used to give Normality, e.g. the logarithmic transformation will by definition transform a log-Normal distribution to a Normal. But there is far too much emphasis on significance tests in current statistical practice (Chapter 6). Those of us who deplore the automatic, unthinking use of significance tests will not be very upset if tests do give inaccurate answers. If those tests are used for their proper purpose—as aids to biological judgement—nothing is lost if the answer is '6 per cent probability' when it should be 5 per cent. Chapter 6 will show that the standard parametric tests (t, F, χ^2) have no firm logical basis, but are best regarded as approximations even when the y-remainders are exactly Normal.

3. As Chapter 4 pointed out, nearly all statistical methods assume that treatments have additive effects—partly because the theory is then algebraically tractable, and partly because the calculations are then straightforward. Suppose we are dealing with percentages—e.g. the proportion of males in a given sample of animals. A treatment which converts 2 per cent to 3 per cent, i.e. adds 1 per cent, is unlikely to change 25 per cent into 26 per cent; more likely, it will increase 25 per cent to perhaps 28 per cent. Since percentages below 0 or above 100 are impossible, the original scale of percentages is tight at either end, but relatively open in the middle. The angular transformation is then used to make the basic assumption, that treatments have additive effects, more reasonable. Similarly, if we are dealing with counts of animals, it is quite likely that some treatment which changes a count from 10 to 20 will be acting multiplicatively, and so change 100 to 200 rather than to 110. The logarithmic transformation then converts the situation to an additive one.

Of reasons (1), (2) and (3), the third is by far the most important. That is because difficulties (1) and (2) may be dealt with in other ways: (1) by using an analysis which is weighted to allow for heterogeneity of variance (Chapter 1) and (2) by working out tests suitable for the particular kind of non-Normal distribution. But since the basic statistical methods assume additivity, we have to use scales of measurement which, at least approximately, make that assumption reasonable. In general, therefore, consideration (3) takes precedence. But reasons (1), (2) and (3) are interconnected, so that a transformation which deals with (3) will quite likely satisfy (1) and (2) as well. That is very generally the case when dealing with continuous measurements like lengths or weights, but not necessarily when dealing with counts. If the size of a treatment effect remains the same at all points on the scale of measurement, it is likely that the size of the relatively small remainders will also be uniform, i.e. that the remainder variance will be the same at all points of the scale. If there is reason to suspect serious heterogeneity of variance, the treatment effects may well be non-additive too. It is therefore more common to use a transformation than a weighted analysis to deal with heterogeneity of variance. If the treatment effects are additive, the relatively small residuals will tend to be additive too, and hence the distribution of the remainders is likely to be approximately Normal (Chapter 4).

If the range of variation in the whole set of data is small, there may be no need to transform, even though the data are of a kind that would otherwise need transformation. For example if we are dealing with percentages varying from 10 to 50, it would be advisable to apply the angular transform; but if the range of percentages is only 20 to 35, we should obtain almost precisely the same answer, whether we transformed or not. That is because a transformation in effect curves the scale of measurement; if the range of observations is small, only a short piece of the scale is used, and the curvature will have no serious effect.

5.2. Some commonly used transformations

5.2.1. The angular (or arc sin $\sqrt{x}$) transformation

This is applied to percentages lying strictly between 0 and 100 (or equivalently to fractions between 0 and 1). As mentioned above, it has the effect of stretching the scale at each end, so it is not suitable where the percentage can exceed 100, e.g. percentage increases in numbers. In such cases we might consider using a logarithmic transformation, which stretches the scale at the bottom end. The angular transform is often applied as a matter of course to binomial or multinomial data, e.g. 'of N animals, so many per cent are of this kind'. Although the angular transform can solve the problem of non-additive treatment effects, it takes no notice of the sample size N. An individual percentage based on N observations will be intrinsically more accurate if N is large than if N is small. So if we are analysing percentages, and if the number N on which the individual percentages are based is very variable, we may need to apply the angular transform to obtain additivity of treatment effects, and then use a weighted analysis to allow for the fact that the different observations, being based on radically different Ns, have different accuracies. In that case the variance of each transformed percentage is proportional to $1/N$, and since the appropriate weight is 1/variance (Chapter 1), the weight used would be simply N.

5.2.2. The square root transformation

This is appropriate to counts with a Poisson distribution. Its effect is to make the larger values less important compared with the smaller. For example, it converts the series 0, 1, 4 to 0, 1, 2. On the original scale, the difference between 0 and 4 is four times that between 0 and 1, whereas on the new scale, the corresponding difference is only twice as large. The greatest value, 4, has become relatively less important. In other words, the transformation stretches the scale of measurement at the bottom end.

5.2.3. The logarithmic transformation

A purely random or Poisson distribution of counts has a variance equal to the mean, so a distribution of counts which has a variance greater than the mean is said to be *over-dispersed*. The logarithmic transformation is usually applied to such distributions, such as the negative Binomial and the Logarithmic. In the case of a truly Negative Binomial distribution, there is a special *ad hoc* transformation which may be used, but it gives much the same answer as the simpler logarithmic transform. The logarithmic transform, like the square-root transform, serves to stretch out the bottom end of the scale; but the effect of the logarithmic is much more severe than that of the square root. To avoid trouble when $n = 0$, it is usual to analyse not $\log(n)$ but $\log(n + c)$ where c is some constant. Usually c is arbitrarily chosen to be 1; the larger the value of c, the less severe is the effect of the transformation. Similarly, $\sqrt{(n + c)}$, where c is a positive constant, is a less severe transform than $\sqrt{n}$. There are theories to prove that for purpose (1), some value of c such as $\frac{3}{8}$ is optimal, but such theories become trivial when it is recognized that the main point of transformations is purpose (3). The logarithmic transform is often used successfully on counts of animal or plant populations, but there is no theoretical reason why it should always be best in such cases.[2]

5.2.4. Probits and logits

These are special-purpose transforms used to facilitate the analysis of biological assays.

Before the advent of computers, it was sometimes difficult to decide which transformation to use in any given case. Fortunately we learn from experience that two transformations of approximately the same severity will give the same biological answer. Indeed, in most cases even a transform as severe as the logarithmic will give the same biological answer as does analysis of the untransformed data. The severity of any given transform may be judged by plotting the transformed scale against the original scale over the range of values concerned in the analysis. In cases where the answer obtained depends on which transformation is used, we must either examine the data to see which transform is most appropriate, or perhaps abandon the data as inadequate to answer the biological question satisfactorily. Nowadays it is a simple matter to perform the same analysis with different transforms to check that the same biological answer is obtained. If the analysis on a logarithmic scale gives much the same answer as analysis on the untransformed scale, it follows that analysis on any intermediate scale would also give the same answer.

The basic principle of statistical analysis (Chapter 1) is to split y_i into two parts, a known function F_i and an unpredictable residual. So far we have considered transforming the values of y itself, i.e. the transformation is applied in blanket fashion to $(F_i + \text{remainder})$. Of the three reasons for transformations given at the start of this chapter, (3) refers to F_i while (1) and (2) concern the remainder. It would therefore be better to apply a transformation to F_i alone for purpose (3) and then to use a method of analysis appropriate to the true distribution of the residuals, whatever it may be. Analyses of this type are known as 'generalized linear models' (Chapter 7). Such analyses can only be done by computer, for the following reason. The division of y_i into a transformed F_i and a remainder can be done only after the transformation has been applied to F_i; but the transformation cannot be applied to F_i until it has been separated from the remainder. The analysis therefore has to be done iteratively, i.e. it requires several cycles of computation to minimize the sum of squares of the residuals—or to maximize the likelihood if more appropriate (Chapter 7). Of course, the program does the whole job automatically once the user has specified what kind of analysis is required. As mentioned earlier in this chapter, a transformation which converts continuous variates (lengths, areas, weights etc.) to additivity will usually ensure approximately Normal residuals too, in which case the regular analysis of transformed ys is perfectly satisfactory. Generalized linear models come into their own for analyses of things like counts, fractions or periods of time before something happens, where transformations to additivity need not produce nearly Normal residuals with uniform variance.

5.3. Presentation of results

Chapter 2 discussed the presentation of numerical results. A special problem may arise when transformations have been used. Means, standard errors, etc. have been estimated on the transformed scale. It is often sufficient to quote the results entirely on the transformed scale, in which case no problem arises; but sometimes the results of an analysis done on the transformed scale have to be translated back to the original scale of measurement. That is easily done for means. For example, if we have analysed a set of original measurements y on the scale of $\log(y)$, we can quote the antilogarithm of the mean value of $\log(y)$. That antilogarithm will in fact be the *geometric* mean of y. The same considerations which prompted us to use the logarithmic transform also assure us that the geometric mean will, in this case, represent the values of y better than a straightforward arithmetic mean. But it is not possible to quote standard errors on the untransformed scale. Standard errors will be symmetrical about the mean on the transformed scale, but not on the original scale. This is merely a technical difficulty: from a biological point of view, if the geometric mean is appropriate, then asymmetric errors (on the untransformed scale) are appropriate too. The difficulty may be avoided in two ways. We may quote, on the original scale, asymmetric confidence limits found by reconverting the symmetric confidence limits of the transformed scale.[3] Alternatively, we can give two sets of results, one with standard errors on the transformed scale, and the corresponding set of means only on the original scale. The reader can then judge both the significance of the results on the transformed scale, and their practical importance on the original scale.

Transformations are also used *mathematically*, to convert a curvilinear relation to an additive one (Chapter 4). It often happens that some transform chosen for mathematical reasons is also suitable statistically. There is no theoretical reason why that should be so, except that linearity and additivity often occur together.

Notes

1. *Derivation of transformation functions.* By definition, the variance V of y is $E(y-m)^2$. The object is to find a function $f(y)$ whose variance is constant, i.e. independent of m. By Taylor's expansion,

$$f(y) = f(m) + (y-m)f'(m)$$

provided that the further terms in $(y-m)^2$ are negligible in comparison. Here $f'(y)$ is the first derivative of $f(y)$, i.e. df/dy. Then the variance of $f(y)$ is

$$E[f(y) - f(m)]^2 = E(y-m)^2 f'^2(m) = Vf'^2(m)$$

If this is a constant c^2, say, $f'(m)$ must be $c/\sqrt{V}$. It is then possible to find the appropriate function $f(y)$ when V is any specified function of m. For example, if the distribution of y is Poisson with mean m, $V=m$ so that $f'(m)$ becomes $c/\sqrt{m}$, and the appropriate transformation $f(y)$ is the square root, $\sqrt{y}$, because that is the function whose first derivative is a constant$/\sqrt{y}$. Or again, if the distribution of y is Binomial with probability p, $V=p(1-p)$ and the appropriate transform of y is the angular transform. Similarly it is possible to derive the appropriate transformation when V is any specified function of the mean. Although the commonly used transformations were derived in this way to stabilize the variance, it is commonly found that they also fulfil the main role of transformations, viz. to achieve additivity.

2. Counts of animals often do not conform to any of the standard distributions. Suppose we are analysing the distribution of the number of tapeworms per wild deer. Quite often there are some animals which have a lot of worms, but many which have none. The distribution may be so lop-sided, or even two-humped, i.e. the frequency of zeros may be so great in comparison with 1s, 2s, 3s, ..., that no transformation can reasonably deal with it. In that case we may have to do separate analyses, the first on the ratio of deer without worms to deer with worms, and the second on the number of worms per infested deer. That is, the zero term is analysed separately. The two analyses will usually reinforce each other, telling the same story.

3. When the transformation is logarithmic, the (additive) standard errors $\pm s$ attached to arithmetic means on the transformed scale of $\log(y)$ convert to multiplicative standard errors $\overset{\times}{\div}$ antilog(s) attached to the corresponding geometric means on the original scale of y, and may be so quoted.

Examples 5

5.1. Plot the following transforms of y, for $y=0, 1, 2, \ldots, 20$:

(a) $\sqrt{y}$ (b) $\log(1+y)$ (c) $\sqrt[3]{y}$ (d) $\exp(y)$
and for $y=0, 5, 10 \ldots 100$
(e) angular transform of y per cent.

Compare the effects and severities of these transforms. It is the *shape* of the transformation that matters, not the absolute size of the transformed values.

5.2. Repeat Example 2.2, using logarithmic and square-root transforms. Convert means back to the original scale of measurement. Do these transformations affect the biological answer in this case?

5.3. The body lengths and skinned carcase weights of twelve rabbits were

Length (cm)	Carcase weight (g)	Length (cm)	Carcase weight (g)
30	2453	25	1358
25	1148	28	1718
26	1406	29	2138
35	3062	23	1051
30	2292	31	2054
32	2981	30	1951

We want to predict carcase weights from body lengths. Why might we expect carcase weight to be proportional to length^3? Plot (a) carcase weight against length, (b) carcase weight against length^3 and (c) $\sqrt[3]{}$(carcase weight) against length. Is it better to use transformation (b) or (c) before doing a linear regression?

5.4. *An extreme case.* Suppose that, of a sample of N animals, all are female. Then the proportion p of males is estimated to be zero, and the (Binomial) variance of that estimate, viz. $p(1-p)/N$, is also zero. So we seem to have arrived at the conclusion that the proportion of males in the population is estimated to be zero with standard error zero, i.e. there are no males at all in the population. The problem arises because we have to use the value of p estimated from the *sample* to calculate the variance, whereas to get the true variance we need to use the unknown *population* value of p. But the absence of males from the sample might be due to chance. No amount of transformation can help us here. Instead we argue as follows. Suppose the true frequency of males is p. The probability that all animals in the sample are female is then $(1-p)^N$. The '95 per cent confidence limit for p' is that value of p which makes $(1-p)^N$ equal to 0.05; any larger value of p would make the all-female sample unlikely. Work out this confidence limit for $N = 10$ and $N = 100$.

6 Significance tests

6.1. The need for common sense

Statistical analyses help the biologist to draw appropriate conclusions—and avoid inappropriate ones—from the data. But the biologist must still take responsibility for those conclusions. Significance tests are only an aid in judging the results of an analysis. The variance ratio, or F-test, is not an integral part of the analysis of variance: it is a tool which can help us interpret the results of that analysis. A test is not the end product of the analysis, but an afterthought which is useless unless it helps us to make sense of the analysis. Some people hope, by the automatic use of significance tests, to avoid personal responsibility for the biological interpretation. That hope is vain. The test says '*either* something improbable has happened *or* there is a real effect'. Purely by chance, the result will be significant at the 5 per cent level once every twenty times; so the test cannot *prove* that there is a real effect. Even if the result of the test is not significant, it doesn't follow that there is no real effect. All that the test can do is assess the chance of obtaining the data in the absence of the effect in question. Significance tests can help in the interpretation of data, but so can common sense, a knowledge of biology, and an intimate acquaintance with the data.

It depends, too, how serious the consequences of mistaken judgement might be. Consider, for example, the rejection of outliers. A set of data sometimes contains an outlying value considerably different from the other values. The outlying value might be genuine or it might be spurious, and in the latter case it ought to be rejected. So we have to decide whether an egregious value may be regarded as part of the sample, or should be rejected as wrong. Inclusion of an erroneous outlier can bias the answer, and—especially—inflate the variance. Sometimes we refer back to the original source of measurement, and sometimes the observation is biological nonsense; otherwise, the question is best settled by common sense. As always, a plot of the data will help. A borderline case will not make much difference to the answer, whether it is included or rejected. There certainly exist special tests to detect outliers, and if applied sufficiently enthusiastically, those tests can eliminate nearly all the observations in the sample!

Particularly serious is a slavish reliance on arbitrary levels of significance. There is nothing magic about 5 per cent and 1 per cent. We shall see later that everyday tests (t, F, etc.) are only approximations anyway, even when the data have perfect Normal distributions. Some people think that if the result is significant at 4 per cent probability it must be accepted, but that if the probability is 6 per cent the result must be ignored. But in fact, both the 4 per cent and the 6 per cent merely assess the possibility that the result occurred by chance: the 6 per cent probability is rather less significant than the 4 per cent. As a general rule, if significance is at this borderline level, further work is needed to establish the reality of the effect in question.

6.2. Types of test

Commonly used are the t-, F- (or variance ratio) and χ^2 tests. They all assume that the remainders are Normally distributed. If that assumption is wrong, the test will wrongly assess the significance probability. If the true distribution were known, the test could be adjusted accordingly. The t-test, which compares means, is more robust (less sensitive to departures from Normality) than the F- and χ^2 tests which are concerned with variances. We shall see that these three tests are interrelated, so that they cannot contradict each other. They can return only one answer to any given question.

Non-parametric tests, on the other hand, avoid the assumption of a Normal (or any other) distribution of remainders. They do so by concentrating on some particular aspect of the data which can be analysed distribution-free. For example, a non-parametric test may consider only the signs of the residuals, not their actual sizes. Non-parametric tests therefore tend to be less efficient than the F-, t- or χ^2 tests when applied to Normally distributed remainders, because they ignore some of the information. But in practice, non-parametric tests compare very favourably with the orthodox tests, especially when the data are not truly Normal.[1] Although non-parametric tests avoid the assumption of Normality, they still require successive observations to be independent (Chapter 4). Sometimes there are several non-parametric tests which may each be used to examine some given question—for instance, the difference between two means. Those tests will give rather different answers, because they concentrate on different aspects of the data. Some biologists try one test after another in the hope that one of them will achieve some arbitrary level of significance. They delude themselves: by insisting on arbitrary 5 per cent or 1 per cent levels, they demonstrate misunderstanding of the nature of significance tests.

The multiple range tests show how the automatic use of significance tests can give misleading results. Suppose we wish to compare a set of treatment means. Chapter 2 mentioned that if the 'treatments' mean square significantly exceeds the remainder mean square, we may take some notice of unexpected differences revealed by inspection of the treatment means. Multiple range tests try to identify the similarities and dissimilarities among the set of treatment means, allowing for the fact that many comparisons (not all independent in the sense of Chapter 2) may be made between them. But it is often the pattern of a set of treatment means, rather than the exact difference between any pair of them, that is important. If the means represent the effects of increasing amounts of fertilizer, they all form a single response curve; it makes no sense to chop them up by individual pairwise comparisons. If further analysis is desired, we might fit a curve (Chapter 13) to summarize the whole pattern of means: but that involves an assumption that the means really should lie on a smooth curve. Many people do not use multiple-range tests at all, but simply examine the whole set of treatment means with their standard errors, bearing in mind their biological meaning. If multiple range tests help that process, well and good; but often they merely delay the time when you have to say 'I think the results mean so-and-so'.

6.2.1. The t-test

A value of t is always 'something, divided by its standard error'. The something might be the difference between a mean and its theoretical value, often zero; or the difference between two means; or perhaps a comparison between regression coefficients. Strictly speaking, a value of t should be compared with the printed table of values of t only when the remainders are Normally distributed. But the t-test is robust, that is, it can tolerate reasonable departures from Normality. And since the test is only an aid to judgement, the precise level of significance is, within reason, unimportant. Here is another way of thinking about the t-test. Its denominator, a standard error, is calculated from a set of remainders— that is, from a set of comparisons among the sample values y. The standard error is therefore a kind of average of many such comparisons. So the value of t is a comparison, divided by the average of many such comparisons. When t is used to test a biological effect which does not in fact exist, the numerator and denominator of t should be about the same size, provided that the standard error has been calculated correctly. It is of course unlikely that they will be exactly the same size. The printed table of t tells us whether the calculated value could reasonably be obtained by chance.

In practice, t must be at least 2 before it can be judged significant. Two rough-and-ready rules can save a lot of time. Suppose we wish to use t to compare two quantities whose remainder variances are V_1 and V_2 respectively. Assuming that the quantities are uncorrelated, the variance of their difference is $V_1 + V_2$, and so the standard error is $\sqrt{(V_1 + V_2)}$. If V_1

and V_2 are roughly equal, $\sqrt{(V_1 + V_2)}$ is roughly 3/2 times either $\sqrt{V_1}$ or $\sqrt{V_2}$; that is, the standard error of the difference is about 3/2 times the standard error of either of the two original quantities. But if V_1 is much bigger than V_2, $\sqrt{(V_1 + V_2)}$ is little bigger than $\sqrt{V_1}$, and so the standard error of the difference is about the same as that of the least accurate of the original quantities. These rules often make it unnecessary to actually calculate values of t except in borderline cases, and even in borderline cases the exact level of significance is not very important.

Since t is 'a comparison, divided by its standard error', it follows that t^2 is 'a comparison^2, divided by its variance'. But the comparison^2 is itself an estimate of variance, estimated from one comparison only and therefore with one degree of freedom. So if t has n degrees of freedom, $t_n^2 = F_{1,n}$, i.e. t^2 is a variance ratio with 1 and n degrees of freedom. That means that any t-test may equally well be done as an F-test. For example, in Chapter 1 the test of male versus female gave a variance ratio of $4.80/1.20 = 4.0$, and in Chapter 2 the test of the same comparison gave $t = 1.60/0.80 = 2.0$, so $t^2 = F$.

6.2.2. The F-test

The variance ratio test is more general than the t-test because F can have more than one degree of freedom in the numerator, but the principle is the same: if there are no true biological effects to swell the numerator, F is the ratio of two independent estimates of the same variance and so is expected to equal 1. As mentioned in Chapter 1, the word 'independent' is important; neither mean square may include the other, or any part of it. As a ridiculous extreme, if we used the same mean square as numerator and denominator, F would necessarily be exactly 1. Since the F-test is concerned with variances, it is not very robust, i.e. it is sensitive to non-Normality—especially skewness—of the remainders (but if the numerator of F has only one degree of freedom, $F = t^2$ and so the F- and t-tests must be equally robust in that case).

Sometimes in a complicated analysis of variance it is difficult to decide which mean square to use as denominator in an F-test. For example, in the two-way table of Chapter 2 we could compare the 'rows' mean square with either the interactions mean square or the residual mean square. But the choice of an appropriate denominator is quite simple. The test is intended to consider whether there might be genuine differences between rows. It compares the actual value of the rows mean square with an estimate of 'what the mean square should be if there were no worthwhile differences between rows'. Then if F is not significantly large, there is no reason to suppose that worthwhile differences exist. So when choosing a denominator for F we need only ask 'How big would we expect the numerator mean square to be, if there are no true biological effects of the kind we are looking for?' Then as denominator we use a residual mean square which estimates the level of variation to be expected, in the absence of such biological effects. It is part of a good experimental design to make sure that an appropriate residual mean square is indeed available (Chapter 8).

Tests of correlation coefficients are another special case of F. Suppose the correlation between two variates y and x, in a sample of size n, is r. If the sum of squares of y (corrected for the mean $\bar{y}$) is S, the analysis of variance of y becomes

	d.f.	s.s.	m.s.
Regression on x	1	$r^2 S$	$r^2 S$
Residual	$n - 2$	$(1 - r^2) S$	$(1 - r^2) S / (n - 2)$
Total (corr. for mean)	$n - 1$	S	

In other words, the regression of y on x absorbs a fraction r^2 of the total sum of squares of y, since r is the amount of linear 'co-relation' between y and x (Chapter 3). Equally, the regression of x on y absorbs a fraction r^2 of the total sum of squares of x. Although the regressions of y on x, and of x on y, are different, the degree of predictability is the same in each case. It follows from this table that the variance ratio for regression/ residual is $(n - 2) r^2 / (1 - r^2)$, which is identical to $F_{1, n-2}$. This identity can be extended to multiple correlations. Incidentally, the F-test does not require that the values of both variates x and y shall be Normally distributed, but only the values of either y or x, depending on which regression is being used (Chapter 3). Since r^2 is equivalent to F by the above identity, the same must be true of r too.

6.2.3. The χ^2 test

The χ^2 test also may be treated as a special case of variance ratio. The following argument shows how. Usually the denominator of F is a residual mean square which estimates the population variance V. Now suppose that the sample includes the whole (infinite) population. The denominator will have infinite degrees of freedom, and it will equal V exactly, because that is how V is defined. This is a purely theoretical argument since it is impossible to take an infinite sample, but sometimes (as in Example 9.1a) we use a theoretical expected value as the denominator of F, and then we say that the denominator, being exact, has infinite degrees of freedom. So in this case F becomes a numerator mean square with n degrees of freedom, divided exactly by V. Unless it is inflated by true biological effects of the kind that are being tested, the numerator mean square will itself equal V on average. But by definition of χ^2, the 'sum of squares/V' has a χ^2 distribution with n degrees of freedom. Therefore the 'mean square/V' has the same distribution as χ_n^2/n. In other words, $F_{n,\infty}$ is the same as χ_n^2/n. So once again the table of χ^2 is superfluous, once we have a table of F. Since χ^2 with n degrees of freedom is expected to equal n, χ_n^2 has to be quite a bit bigger than n before it can be significant.

The χ^2 distribution is very important in statistical theory, because it describes the behaviour of sums of squares of Normal deviates, and so underlies the whole theory of sums of squares, standard errors etc. χ^2 has the important property of additivity. If y_i is Normal with mean m_i and variance V_i, $(y_i - m_i)^2/V_i$ is χ^2 with one degree of freedom. Its average value is 1, because V_i is defined as the average value of $(y_i - m_i)^2$. The sum of n such values, $\sum(y_i - m_i)^2/V_i$ is χ^2 with n degrees of freedom. Its average value is n. In other words, χ_n^2 is the sum of n independent values of χ_1^2; and the sum of χ^2 with n_1 degrees of freedom and an independent χ^2 with n_2 degrees of freedom is χ^2 with $(n_1 + n_2)$ degrees of freedom. This is the familiar property that you can add together sums of squares or conversely, analyse a sum of squares into independent parts.

χ^2 is mostly used in practice to analyse counts, i.e. frequency distributions, contingency tables etc. The usual expressions for χ^2 are really weighted sums of squares (Chapter 1). But counts are not distributed Normally. They must be whole numbers. Very often it is good enough to treat the count *residuals* as approximately Normal, even though the original counts are not. The calculated values of χ^2 then conform approximately to the theoretical distribution of χ^2, but we must watch out for cases where the approximation breaks down. This is likely to happen when the average or expected count is less than five. Such cases may be treated by the likelihood-ratio method of Chapter 7.

Many pitfalls await us when we use χ^2 to analyse contingency tables. To be sure that the χ^2 approximation can be trusted, we require the expected (not the observed) value for each cell to be at least five. The entries in a contingency table must be independent and complete. Suppose we take blood samples from 100 humans, of whom 47 are male. 31 are found to be Rhesus negative, of whom 17 are male. It would be wrong to construct the following table:

	Humans	Males
Rh−	31	17
Rh+	69	30
Total	100	47

because the humans column *includes* the males column. It is essential to split the table into independent entries:

	Females	Males	Total
Rh $-$	14	17	31
Rh $+$	39	30	69
Total	53	47	100

The table must also be *complete*. Suppose we wish to test the sex ratio for 1:1. It is wrong to say 'of 100 people, 50 males were expected but 47 were observed, so $(47-50)^2/50$ is χ^2 with one degree of freedom'. It is essential to include the females as well:

Males	Females
47	53
(50 expected)	(50 expected)

So χ^2 is $(47-50)^2/50 + (53-50)^2/50$. It has only one degree of freedom because the two expected values, each 50, must add up to the grand total for the sample, 100.[2]

Another difficulty concerns the combination of different contingency tables. For example, the two tables

80	20
20	5

and

10	10
40	40

reveal exact proportionality, the χ^2 (1 degree of freedom) for heterogeneity, i.e. disproportionality, being zero in both cases. But if we injudiciously add the two tables together, we get a table which is very disproportionate:

90	30
60	45

This example shows that individual contingency tables may not be added together, except in particular circumstances. In this case, the heterogeneity of the combined table consists solely of heterogeneity *between* the two original tables, whereas the biologist is presumably only interested in heterogeneity within those tables. One way of combining the evidence from several tables is to calculate a value of χ^2 for each table and add together those values of χ^2 to get an overall χ^2. This method is often unsatisfactory because χ^2 takes no account of the *direction* of the heterogeneity. For example, the two tables

	Males	Females
Rh $-$	17	14
Rh $+$	30	39

and

	Males	Females
Rh $-$	14	17
Rh $+$	39	30

yield the same value of χ^2, but the first shows a slight excess of Rh negative males and Rh positive females, while the second shows the reverse. When combined, two such tables ought to cancel, whereas two tables with deviations in the same direction ought to reinforce each other. Yates (1955) gives a better way of combining a set of 2×2 tables which does take account of direction (Example 6.2).

Values of χ^2, calculated from frequency data, only approximate the true χ^2 distribution. Yates's correction for continuity improves the approximation. It may only be used when χ^2 has one degree of freedom, for example in a 2×2 contingency table. Yates's correction should not be used when calculating values of χ^2 which will themselves be used in further calculations, e.g. added to other values of χ^2. It should be used only when the value of χ^2 will be used immediately to assess significance. In effect, Yates's correction calculates a wrong value of χ^2 which gives a correct estimate of the level of significance.

Although χ^2 gives a good *test* of heterogeneity, it is a poor *measure*. The two tables

90	30		900	300
60	45	and	600	450

show precisely the same degree of disproportion, yet χ^2 in the second is ten times larger than in the first.

6.3. Concluding comments on significance tests

Sometimes a value of F, t or χ^2 is significantly *small* (say, $P > 95\%$). Such values are just as remarkable in their way as significantly large ones: they will of course occur by chance once in twenty times. But if genuine, this result implies either that the test has been misapplied, or the data have somehow been selected or adjusted to give a suspiciously exact answer, or some biological mechanism is acting to control variation.

The common parametric tests are all special cases of variance ratio or F. In the null case when there are no genuine biological effects to swell the numerator mean square, F is expected to be 1. If both the numerator and the denominator mean squares had infinite degrees of freedom, i.e. were exact measures of V, F would necessarily be exactly 1. In the printed tables of critical values of F, the value for $n_1 = n_2 = \infty$ is always 1.00, no matter what the level of significance probability. That is because any departure from 1, however slight, would mean that the numerator and denominator variances could not be equal. More generally, F is the ratio of two mean squares with n_1 and n_2 degrees of freedom, each estimating the same variance V. As n_1 increases, the numerator becomes a more accurate estimate of V, and so the critical value of F (the value which must be exceeded, to establish significance) must decrease towards 1.00. Similarly for n_2: if either n_1 or n_2 increases, the calculated value of F becomes more accurate and the corresponding critical value of F, used to assess significance, must logically decrease. And so, in general, it does. If you look at the standard printed tables of variance ratio (Fisher and Yates 1963) which appear in most statistical textbooks but not this one, you will see that F does always decrease down the columns as n_2 increases. And in general, F decreases across the rows as n_1 increases—but not always. When $n_2 = 1$ or 2, the values of F actually increase as n_1 increases, which is logically impossible. The values are correctly calculated, so there must be a logical blunder somewhere.

Significance tests generally assess the probability, not just of the observed value of the test statistic (F, t, etc.) but also of more extreme values which have not been observed. Using an elegant and convincing argument, Jeffreys (1939) showed that it is not correct to include the more extreme values: we should consider only the value actually observed. Fisher (1956) accepted Jeffreys' argument. It is fair comment to say that Fisher would never have done so if he could have seen any way round the argument! In the case of discrete distributions, e.g. contingency tables with only one degree of freedom, the argument means that we should consider only the probability of the value actually observed, not of the more extreme values too. That makes little difference, since the combined probability of the more extreme values is always small compared to the probability of the observed value itself. But usually we are dealing with measurements which have continuous distributions, in which case the probability of the value actually observed must be infinitesimal. The basic idea of the customary significance test— to assess the probability of obtaining the observed value by chance—then breaks down completely, since the probability of obtaining that exact value, and no other, is effectively zero. Fisher recommended that the regular tests, based on probability, should be replaced by 'likelihood ratio' tests considered in Chapter 7.

This argument means that the regular tests—t, χ^2, F—are logically defective. So why do they work as well as they do? Because their test statistics are 1:1 functions of 'likelihood ratio'. In other words, the F-test is an approximation to the corresponding likelihood-ratio test. Both tests ask precisely the same question; the only difference lies in just how they examine it. Here the argument comes full circle. It is possible to calculate a table of critical points of F which correspond, not to fixed probabilities such as 5 per cent, but to fixed values of the likelihood ratio. Those values of F always decrease towards 1.00 as either n_1 or n_2 increases: they are not open to the objection mentioned above. This discussion underlines the futility of slavish adherence to 5 per cent or 1 per cent probability levels—the significance probabilities generated by the standard tests are not quite what we want anyway. There is no reason why we should not continue to use such tests as aids to judgement, since they work well enough in practice. In any case, their assumptions of Normality and perfect independence can never be completely verified from samples of finite size.

Notes

1. There is one case, rarely encountered, where the non-parametric test is greatly superior to its parametric counterpart. The χ^2 goodness-of-fit test considers only the sizes of the successive deviations from a theoretical distribution or curve, whereas the non-parametric Kolmogorov–Smirnov test examines their signs, and is therefore capable of detecting small but consistent deviations of consecutive data points from the curve. These remarks apply only when the data are tested against some theoretical distribution or curve which is specified entirely *a priori*. If, as is usually the case, the parameters of the distribution or curve have to be estimated from the data themselves, difficulties arise and the Kolmogorov test ceases to be non-parametric. The same shortcoming of χ^2—that it does not consider whether different residuals have the same, or opposite, signs—recurs later in the chapter in a different context, viz. the combination of a set of 2×2 contingency tables.

2. Consider a sample of size N from a Binomial distribution with probabilities p and q, where $p + q = 1$. The numbers observed in the sample are n_1 and n_2, where $n_1 + n_2 = N$. The expected value of n_1 is Np, and of n_2 is Nq. If $n_1 = Np + d$, n_2 must equal $Nq - d$. The value of χ^2 with one degree of freedom, viz.

$$(n_1 - Np)^2/Np + (n_2 - Nq)^2/Nq,$$

becomes $d^2/Np + d^2/Nq$. This value of χ^2 is calculated assuming that n_1 has variance Np and n_2 has variance Nq, i.e. as if n_1 and n_2 were Poisson variates with means Np and Nq: but then χ^2 has only one degree of freedom, not two, because the expected values Np and Nq must add up to the sample size N. But $d^2/Np + d^2/Nq$ is the same as d^2/Npq, which is χ^2 with one degree of freedom derived from the Binomial deviation d with variance Npq. So there are two ways of calculating χ^2: either as the sum of two terms, one for n_1 and one for n_2, derived from the Poisson distribution but recognizing that Np and Nq are constrained to total $N = n_1 + n_2$; or by treating the deviation d itself as Binomial. The answer obtained by either method is the same—as it must be.

Examples 6

6.1. (a) From printed tables of t, F and χ^2, verify that $t_n^2 = F_{1,n}$ and that $\chi_n^2/n = F_{n,\infty}$.

(b) Suppose that $\bar{y}_1$ and $\bar{y}_2$ are the means of two samples, each of size N, with combined residual mean square s^2. Show that to test the difference between $\bar{y}_1$ and $\bar{y}_2$,

$$t = (\bar{y}_1 - \bar{y}_2)/\sqrt{(2s^2/N)} \qquad \text{and} \qquad F = N(\bar{y}_1 - \bar{y}_2)^2/2s^2,$$

so that $t^2 = F$. This example confirms the statement in Chapter 1 that the 'between' mean square is indeed concerned with the average differences between the categories or treatments.

6.2. *Combination of 2×2 contingency tables.* In samples of students in Australia and Canada, the numbers of horse owners were

	Australia		Canada	
	Men	Women	Men	Women
Horse	14	23	4	11
No horse	14	9	29	19

Are men and women equally likely to own horses? Show that in the Australian table $\chi_1^2 = 3.023$ and in the Canadian, $\chi_1^2 = 5.219$. These values are calculated without Yates' correction for continuity because they will be used in further calculations.

(a) Adding together the two values of χ^2, we get $\chi_2^2 = 8.242$. This takes no account of the direction of the deviations, i.e. of the fact that in both tables there is an excess of women horseowners.

(b) *Yates's (1955) method.* χ^2 with one degree of freedom is by definition the square of a Normal deviate with zero mean and unit variance. Therefore, $\sqrt{3.023} = +1.739$ may be treated as Normal ($m = 0$, $V = 1$). Similarly $\sqrt{5.219} = +2.285$ is Normal (0, 1). Both values are given the same sign to show that the deviation is in the same direction in both tables. If the deviations were in opposite directions, one sign (it doesn't matter which) would be taken positive and the other negative. The sum $+1.739 + 2.285 = 4.024$ is therefore Normal (0, 2) because the sum of two or more Normal variates is itself a Normal variate. $4.024/\sqrt{2} = 2.846$ is therefore Normal (0, 1). There are several ways of testing 2.846 as a Normal (0, 1) deviate. It can be compared with a table of the Normal distribution, or equivalently with a table of t with infinite degrees of freedom. Alternatively, 2.846^2 is χ^2 with one degree of freedom, which is equivalent to $F_{1,\infty}$.

Compare the results of (a) and (b). This method may be used to combine any number of 2×2 tables. Just like χ^2 itself, Yates's method gives a good test of the existence of a consistent effect, but a bad measure of the size of the effect.

7 Some underlying theory

This is a difficult chapter. It digs into the logical principles which underlie everyday statistical methods. Rarely do we apply those principles directly, when deciding how to analyse any given case in practice; but some acquaintance, however slight, with principles can help to prevent serious mistakes. It very soon appears that statistical methods are built on mud. This does not matter in practice, because the mud is strong enough to bear the structure of everyday statistical methods. Statisticians agree well enough on what to do, but argue fiercely about why they do it. Statistical methods work in practice, but nobody has yet found a watertight justification for making inferences. The same difficulty besets not just statistics, but all scientific research and indeed everyday life. If we knew a foolproof logical justification for making inferences of any kind, we could apply it to statistical inference in particular, and vice versa. This chapter will consider the statistical aspect of the wider problem.

7.1. Bayes's Theorem

Suppose we have a sample S which gives us some information about a parameter θ. Perhaps θ might be a population mean or a regression coefficient. Before the sample S is taken, we do not necessarily consider that all possible values of θ are equally probable. So we say that there is a prior probability $p(\theta_i)$ that θ shall equal any given value θ_i. Some people regard $p(\theta_i)$ as a personal assessment peculiar to themselves, of all the possible values which θ can take; others regard $p(\theta_i)$ as reflecting the structure of the real world. Having obtained the sample S, we may use its information to update our assessment of θ, or in other words, to convert the prior probabilities $p(\theta_i)$ into new probabilities which take account of the sample S. That is done by Bayes' Theorem, which says that the probability (in the light of S) that $\theta = \theta_i$, is proportional to $p(\theta_i)L(S|\theta_i)$. Here $L(S|\theta_i)$ is the probability of obtaining the sample S when $\theta = \theta_i$. ('$S|\theta_i$' does not mean 'S divided by θ_i' but 'S, given that $\theta = \theta_i$'.) $L(S|\theta_i)$ is known as the likelihood of obtaining S when $\theta = \theta_i$; we shall use it to help us compare the different possible values of θ in the light of the observed sample S.

Bayes's Theorem tells us to multiply the likelihood by the prior probability, in order to obtain a new assessment of the probability that

$\theta = \theta_i$. It is here that the trouble starts. Nobody denies that Bayes's Theorem is algebraically correct. The disputes arise over its application. The likelihood $L(S|\theta_i)$ is perfectly well defined, and its value may be calculated for any S and θ_i. It is obtained from the probabilities of the individual observations which make up the sample S. But the theorem asserts that, before we can use $L(S|\theta_i)$ to compare the probabilities of different values of θ_i in the light of S, we must multiply $L(S|\theta_i)$ by the prior probability $p(\theta_i)$. We must therefore assign values to $p(\theta_i)$ for all values of θ_i.

7.1.1. The question of prior probabilities

Here there arises the dispute between 'Bayesians' and 'non-Bayesians'. Bayesians tend to be mathematical statisticians with tidy minds, who like to keep things as logical as possible. For them, Bayes's Theorem is the only possible justification for making inferences, and the practical difficulties of assigning values to $p(\theta_i)$ are of secondary importance. Using Bayes's Theorem, it is never possible to arrive at a probability of 1 for any proposition whatever, i.e. we can never achieve complete certainty about anything. Non-Bayesians tend to be practical statisticians for whom everyday problems of implementation are more important than theoretical nicety. For them, it's no use having a theorem unless the required values of $p(\theta_i)$ can be supplied in some unequivocal way. Both views evidently have merit.

It could be that some previous experience (i.e. some previous sample) gives information about θ; but Bayes's Theorem says that we cannot use that experience to give values of $p(\theta_i)$ without the aid of some initial set of prior probabilities. However much experience we may have, the theorem requires us to return in time to a starting point where we must assign prior probabilities $p(\theta_i)$ to different values of θ_i when we are completely ignorant about θ. So a plea of 'previous experience' cannot extricate us from the dilemma—what is the value of $p(\theta_i)$ when we know nothing about θ? This dilemma is not very serious in practice. Suppose we are comparing two possible values of θ. As evidence accumulates, i.e. as the size of the sample S increases, the comparison comes to depend more and more on the likelihoods, which soon outweigh the prior probabilities. So in practice it doesn't matter much what values of $p(\theta_i)$ we adopt, provided that they are not ridiculously small or zero—that is, provided we admit that θ_i is a possible value of θ. But to justify the whole process of statistical inference we must, according to Bayes, find some way of assigning values to $p(\theta_i)$ when we know nothing about θ. In the absence of an agreed way of doing so, Bayesian methods must appear somewhat arbitrary—but they certainly win on logic.

Some people argue that when we know nothing about θ, all values of θ

are equally possible, and so all values of $p(\theta_i)$ must be the same. There are many objections. For example, the argument seems to contradict itself. If $\phi = 1/\theta$, a uniform distribution of $p(\theta_i)$ implies a non-uniform distribution of $p(\phi_i)$: but we know nothing of ϕ, just as we know nothing of θ, so the distribution of $p(\phi_i)$ ought to be uniform too. But the greatest objection is that many people find the argument incredible. They flatly deny that because we know nothing about θ, we can say that all possible values of θ are equally probable. If we consider $p(\theta_i)$ to be a personal subjective assessment, different people may validly use different values for $p(\theta_i)$, in which case anyone may, at least in theory, validly draw any conclusion he chooses from a given set of evidence. If we consider $p(\theta_i)$ to be some kind of objective probability reflecting the real world, the values of $p(\theta_i)$ must be the same for everybody; but people disagree about the values to be used in practice, when everyone is equally ignorant of θ. It may be argued that we are never completely ignorant about any possible question. Even when we have no directly relevant evidence, total experience can always give us some idea of what to expect. But vague experience cannot be turned into the precise numerical probabilities which Bayes's Theorem requires. To sum up, there is no indisputable way of assessing prior probabilities, but in practice any reasonable set of prior probabilities will do.

Another school of thought, that of Fisher, admits the algebraic validity of Bayes's Theorem but denies its relevance to problems of statistical inference. The value of $p(\theta_i)$ is unknowable, and therefore cannot be used in practice. Some people go further, and deny that $p(\theta_i)$ exists at all. Instead we remember that the likelihood $L(S|\theta_i)$ is in fact the likelihood of obtaining the observed sample S, given that $\theta = \theta_i$. Since the likelihood sums up all that we know about θ_i, we need only consider the likelihood itself. In particular, we take that value of θ which maximizes the likelihood to be the 'best' estimate of θ.

It is obvious that maximizing $L(S|\theta_i)$ would be the same as maximizing $p(\theta_i)L(S|\theta_i)$ if $p(\theta_i)$ were the same for all values of θ_i. So Bayesians say that Fisher's disciples delude themselves—the principle of maximum likelihood really uses Bayes's Theorem, with a hidden assumption that $p(\theta_i)$ is the same for all values of θ_i. Fisher's school replies that when we maximize the likelihood, we maximize the likelihood and nothing else. To accept the likelihood as supreme arbiter is a new basic principle. However gratifying its consequences may be, the principle itself has to be taken on trust.[1]

7.1.2. Sufficiency

There are still some technical questions which arise, whether or not we use prior probabilities. They concern the idea of 'sufficiency'. A sufficient estimate of θ, calculated from the values in the sample S, is one that absorbs all the information about θ contained in the sample. For example, if the values of y in the sample are Normally distributed, the sample can tell us nothing more about the population mean m, once we know the arithmetic mean $\bar{y}$ of the sample. The sample mean $\bar{y}$ is a sufficient estimate of m. That is a very important idea, since it means that once we have analysed a set of data in an appropriate way, it is useless to try alternative analyses in the hope of getting more information. In this case, once we know the arithmetic mean of the sample, we need not waste time trying to extract any further information about the population mean. By contrast, an alternative estimate of m, viz. the average of the smallest and largest values in the sample, would not be 'sufficient' and could certainly be improved on.

Likelihood theory shows that in those cases where it is possible to find a single sufficient estimate at all, the maximum likelihood estimate (i.e. the value of θ which maximizes the likelihood) will itself be sufficient. Whenever the distribution of y permits a sufficient estimate of θ, there is no problem; we use the maximum likelihood estimate or some equivalent which may be more convenient to use in practice, confident that we cannot do better. But for some distributions of y, no sufficient estimate of θ exists. In that case, although the likelihood still sums up all that the sample tells us about θ, no one estimate of θ can embody all that information. Fortunately, if y follows the basic distributions—Binomial and Poisson for counts, Normal for continuous measurements—the everyday estimates of means, regressions etc. are sufficient.

As an example we return to the equation $y_i = F_i + \text{remainder}$, used in Chapter 1. Suppose that the remainder is Normally distributed with variance V_i. That means that the probability of obtaining a particular value y_i is proportional to

$$\exp[-(y_i - F_i)^2/2V_i].$$

The likelihood of a sample of N independent values of y is proportional to the product of the N individual probabilities, and so it is proportional to

$$\exp[-\sum_{}^{N}(y_i - F_i)^2/2V_i].$$

We remember that F_i might represent a mean, the same for all the ys, or perhaps a regression on some concomitant observation x_i. In either case, to estimate F_i by maximum likelihood, we need to maximize

$$\exp[-\sum(y_i - F_i)^2/2V_i]$$

which is equivalent to minimizing the weighted sum of squares $\sum(y_i - F_i)^2/V_i$. Usually V_i is the same for every y_i, in which case we minimize the unweighted sum of squares $\sum(y_i - F_i)^2$. So 'least squares' may be regarded either as a principle in its own right, regardless of how y is distributed, or as a special case of maximum likelihood which arises when the remainders are Normally distributed.

7.2. Generalized linear models

Chapter 5 mentioned 'generalized linear models'. McCullagh and Nelder (1983) describe this subject in mathematical detail.[2] Such analyses, which are particularly useful in the case of contingency tables, illustrate the use of transformations and of likelihood. Consider a two-way contingency table, such as the numbers of Rhesus negatives and positives among males and females (Chapter 6). When there is no association between sex and blood group, the frequencies in the table will be proportional. If there are twice as many negatives as positives among males, the same will be true of females. Consequently we expect that if n_{ij} is the number observed in the (i, j) cell,

$$n_{ij} = a_i b_j + \text{residual}. \tag{7.1}$$

In other words, $F_{ij} = a_i \times b_j$ where a_i and b_j are the appropriate (unknown) row and column factors for that cell. The logarithmic transformation will convert to additivity; $\log(F_{ij}) = \log(a_i) + \log(b_j)$. But it is not very appropriate to analyse the values of $\log(n_{ij})$ as a two-way table of continuous measurements, as in Chapter 2, because the analysis in Chapter 2 assumed that all the ys had the same variance. That is not true of counts or, in general, of log(counts). So the generalized linear analysis applies the logarithmic transformation to F_{ij} to achieve additivity, and simultaneously maximizes the likelihood of the n_{ij}s, regarded as Poisson or Binomial counts (Note 2, Chapter 6). In practice the computer may use, not the likelihood itself, but a 'quasi-likelihood' which is equivalent to the sum of squares weighted inversely by the variance (Note 1, Chapter 2) and which gives equally satisfactory results. If the simple multiplicative eqn (7.1) does not completely describe the data, i.e. if $\log(F_{ij})$ is not perfectly additive, it is possible to insert interactions which are precisely analogous to the interactions of the two-way tables in Chapter 2. Such generalized linear analyses can only be done by computer. They provide deeper penetration, especially in analyses of contingency tables. Although the analyses themselves are perfectly satisfactory, it is difficult to find valid significance tests. The use of a non-additive model such as (7.1) invalidates the regular tests (Note 1, Chapter 4) even though F_{ij} can be transformed to additivity.

Now we can put together various points discussed in this book. Transformed scales of measurement are used to make the effects of treatments approximately additive, partly because we can think most easily in additive terms, and partly because statistical analyses are based on additive models. If the treatment effects are additive, it is quite likely that the remainders will be nearly Normally distributed. Likelihood theory then justifies the use of least-squares estimates. In cases where the residual variance is not uniform, the sum of squares is weighted inversely by the variance. The least-squares estimates will be sufficient and in that sense, optimal. Least-squares estimates of means, regression coefficients etc. are robust, so that the biological answer is not seriously affected by reasonable departures from Normality. All is well, provided that we can accept likelihood theory. That means that we must either arbitrarily accept the likelihood as supreme arbiter—a role which it usually performs very well—or use arbitrary prior probabilities in Bayes' Theorem.

7.3. Significance tests

Suppose that a factory, mass-producing some product, needs to test each batch to check the quality of the articles. Very often it is too expensive to test every item, so the factory takes a sample from each batch. If the sample is satisfactory, the batch is accepted; if not, it is rejected. A firm decision to accept or reject has to be made for every batch. Neyman and Pearson developed the theory of significance tests along analogous lines. A significance test, they say, must either accept or reject a null hypothesis (which might be, for example, that the population mean m is zero). The theory concerns the chances of making wrong decisions—of rejecting the null hypothesis when it is true, or of accepting it when it is false. Many statisticians accept the theory, but others agree with Fisher, who said that it is a false analogy. In scientific research, significance tests are not used to accept or reject hypotheses, but only to assess the weight of evidence for or against them. We can never finally reject any hypothesis, however improbable it may become. Therefore, according to Fisher, analogy with decision making is not pertinent.

Now let's consider what we do in practice. If the significance probability is large, say greater than 1 per cent, we do not actually reject the null hypothesis but assess it as unlikely. But if the probability is small, 1 in 1000 or less, we jump to certainty and decisively reject the null hypothesis. Such behaviour cannot be logical, but is certainly rational. If you are being charged by an elephant, you should logically stop to assess the probabilities of all possible outcomes—for example, that the elephant might drop dead of heart failure—before deciding what to do, but in practice it would be advantageous to ignore improbable outcomes. So Fisher's point of view is right when the significance is slight, but Neyman–Pearson theory is applicable when the significance is strong. We cannot say exactly where the transition from assessment to rejection occurs. It depends in any case on the practical consequences of making a mistake, but the transition is made somewhere in between the customary arbitrary levels of 5 per cent, 1 per cent and 0.1 per cent. That is why those standard levels were chosen in the first place.[3]

As mentioned in Chapter 6, the customary significance tests—t, F, etc.—are logically defective, although they work well enough in practice. Fisher (1956) proposed, in place of the usual type of significance test which assesses a probability, to use the likelihood ratio as criterion instead. We have already noted that the likelihood $L(S|\theta)$ embodies the information conveyed by the sample S about the unknown parameter θ. So if the likelihood decreases rapidly on either side of its maximum, we need only consider values of θ close to the maximum. Likelihood theory uses the rate of that decrease to judge the accuracy of the estimate of θ—and in particular, to estimate its standard error.

Suppose that somebody proposes for our consideration a null hypothesis that the true value of θ, in the population from which the sample S was drawn, is θ_0. To assess that hypothesis we may compare $L(S|\theta)$ when θ takes its maximum-likelihood value, i.e. the maximum value of $L(S|\theta)$, with $L(S|\theta)$ when $\theta = \theta_0$. The ratio of the two likelihoods cannot be less than 1, simply because θ is chosen to maximize $L(S|\theta)$. If the ratio is large, we shall suppose that the null hypothesis (that $\theta = \theta_0$) is unlikely, because alternative values of θ are much more likely. This significance test makes no reference to probabilities. Instead of arbitrary 5 per cent or 1 per cent probability levels, it substitutes arbitrary likelihood-ratio levels. Although this type of test avoids many of the difficulties which beset orthodox significance tests, it is not yet in general use.[4] (Some existing tests convert the likelihood ratio into a corresponding probability, thereby exposing themselves to Jeffreys' (1939) criticism.) Bayesians will not accept this likelihood-ratio test as it stands; for them, the two likelihoods compared must first be multiplied by their appropriate prior probabilities.

This chapter has briefly examined some of the logical difficulties which beset statistical methods. The difficulties are not too serious, because the various approaches all lead to essentially the same practical recipes. Although orthodox significance tests may be a bit dubious logically, they work well enough in practice. But the absence of a watertight logical basis warns us that statistical methods are not foolproof, but must be used with common sense.

Notes

1. The method of maximum likelihood is itself not consistent in the following sense, when applied to samples of finite size. Consider a sample of N values y_i

from a Normal population with mean m and variance V. The log likelihood (using natural logarithms) is given by

$$-2\log(L) = N\log(V) + \sum_{i}^{N}(y_i - m)^2/V. \tag{7.2}$$

This expression for the log likelihood is not an approximation, but is exact. The object now is to estimate V. Before doing so, we must get rid of the unknown nuisance parameter m by replacing it with its maximum likelihood estimator $\bar{y}$. Then (7.2) becomes

$$-2\log(L) = N\log(V) + \sum_{i}^{N}(y_i - \bar{y})^2/V. \tag{7.3}$$

But (7.3) is no longer exact, because $\sum_{i}^{N}(y_i - \bar{y})^2/V$ has a χ^2 distribution with $(N-1)$ degrees of freedom, so that its exact likelihood is given by

$$-2\log(L) = (N-1)\log(V) + \sum_{i}^{N}(y_i - \bar{y})^2/V. \tag{7.4}$$

So the application of maximum likelihood to the exact likelihood (7.2) produces, not the exact (7.4), but the inexact (7.3). This is a useful reminder that although likelihood methods generally work well for samples of finite size, practical difficulties do arise in some cases.

2. Healy (1988) describes how to use GLIM, a specially written computer program for generalized linear models. The same program could also be used for all the worked examples in this book—which illustrates the essential unity of statistical methods.

3. This line of reasoning may be extended to the problem of inference in general, whether statistical or otherwise. The basic problem is 'Why suppose that the sun will rise tomorrow, given that it has risen every morning in recorded history?' (To object that if the sun didn't rise, it wouldn't *be* tomorrow, is to evade the logical problem!) The argument that 'we make inferences because the inferential method has worked well in the past' is logically circular because it uses an inference to justify inference. But we would not make inferences if they had proved generally unsuccessful in the past.

In practice, the human mind jumps to certainty when the probability is close enough to one. We do not just think it highly probable that the sun will rise tomorrow; we take it for granted that it will. The jump to certainty, although illogical, is rational because advantageous in practice. It occurs somewhere between 95 and 99.9 per cent probability, depending on the seriousness of the consequences of being mistaken. There is some empirical evidence on this aspect of human behaviour (*Science* **234**, 542 (1986)).

We have made this same jump to certainty about the process of inference itself: inferences are justified because the process of inference has worked so often in the past that humans are certain it will work again. It is evident that inference would not work unless there were a principle of continuity in the universe, i.e. that things will continue to work in much the same way as before, at least on the human time-scale. Without such continuity, human life itself would be biologically impossible. Yet the various theories of statistical inference make no explicit reference to any such principle of continuity.

4. Such likelihood ratio tests suffer from an unresolved limitation. A likelihood ratio, maximized over two or more parameters simultaneously, will naturally

come out larger than a likelihood ratio maximized for one parameter only. At present there is no valid way of saying what the critical levels should be for likelihood ratios maximized over more than one parameter, corresponding to the arbitrary critical levels adopted for likelihood ratios maximized over single parameters. In other words, we can use likelihood ratio to test only one parameter at a time. We cannot test multiple hypotheses except by dissecting them into a series of simple (one-parameter) hypotheses. This is more an irritating theoretical deficiency than a practical limitation.

Examples 7

7.1. An insect parasite lays one egg in each of 100 larvae. Her progeny include 36 males and 64 females.
(a) Use χ^2 to test for 1:1 sex ratio.
(b) Suppose the true (population) proportion of females is p. The observed fraction $64/100$ is an estimate of p. In a sample of N individuals, the variance of the estimate is $p(1-p)/N$. Use this variance to find the standard error of the estimate for $N = 100$ and $p = 0.64$. Since this standard error is derived from theory, it has infinite degrees of freedom. Use $t = 1.96$ to find 95 per cent confidence limits for p.
(c) Find the angular transform (degrees) of the proportion of females. The variance of the angular transform is $820.7/N$, whatever the original value of p (Fisher and Yates 1963, Table XII). The angular transform is in fact derived according to the argument of Note 1, Chapter 5 to stabilize the variance $p(1-p)$. Find 95 per cent confidence limits on the transformed scale and convert them back to the original scale.
(d) Here is the likelihood treatment of the same problem. The probability of obtaining one female is p. The probability of independently obtaining 64 females is therefore the product of 64 ps, i.e. p^{64}. Similarly for the males. The likelihood $L(S|p)$ is therefore proportional to $p^{64}(1-p)^{36}$, and $\log(L)$ is $64 \log(p) + 36 \log(1-p)$. Calculate $\log(L)$ for $p = 0.52$, 0.56, 0.60, 0.64, 0.68, 0.72 and 0.76. (You can if you wish read the values of p as a variate into the computer and make it calculate $\log(L)$ as a new variate.) By plotting $\log(L)$ against p, or by calculus, show that the maximum likelihood occurs when $p = 0.64$, which is therefore the maximum likelihood estimate of the proportion of females. It equals the proportion of females in the sample. Calculate the likelihood ratio, i.e. [the maximum $L(S|p=0.64)/L(S|p)$] for $p = 0.52$, 0.56, $\ldots$, 0.76 and plot it against p. Read off the values of p for which the likelihood ratio takes the arbitrary value 6. Are those values of p (i) close to, (ii) the same as, the 95 per cent confidence limits in (c)?†

†If n_1 is the observed number of females and n_2 the number of males, $\log(L)$ is $n_1\log(p) + n_2\log(1-p)$. That is exactly true whatever n_1 and n_2 may be. The *expected* values of n_1 and n_2 are Np and $N(1-p)$. The *expected* value of $\log(L)$ is therefore proportional to

$$N[p\log(p) + (1-p)\log(1-p)].$$

The expression in brackets is the 'information' (Chapter 9). Information theory is therefore an approximate or large-scale version of likelihood theory.

7.2. In Example 7.1 the value of p was in the middle of the range from 0 to 1, and the various methods all gave similar answers. But now repeat Example 7.1 for a sample containing 97 females and 3 males. Try values of $p = 0.92$, 0.93, 0.96, 0.97, 0.98 and 0.99. In this case p is so close to 1.0 that the confidence limits in (b) extend beyond 1.0, but methods (c) and (d) allow for the tightness of the scale.

8 Experiments and samples

It is a sad fact that, even now, statisticians are sometimes asked to analyse the results of experiments which are so badly designed that no conclusions can validly be drawn from them. That is why statisticians ask to be consulted before the experiment is done. This does not mean that statisticians are superior beings. It means that experimental design involves both statistical principles, and practical considerations peculiar to the field of enquiry. This chapter will look at the principles, but practical designs must also consider the practical limitations:

Quite commonly it will be found that convenient experimental techniques come into conflict with statistical requirements especially perhaps that of randomization. Educated judgement is required in such cases, the answer is not always an insistence on the strict letter of the laws, rather that we should perhaps require an appreciation of the risks of deviations from them.

The quotation comes from Sprent (1970), an amusing article written in non-technical language which can tell you how the world looks to a highly empirical statistician. For example, if we want to do an experiment on elephants and have only two elephants, should we apply a different treatment to each and hope for the best? Or use the same animals several times? Or abandon the idea until more elephants are available? There are many books devoted to experimental design, most of them written for statisticians. But in Pearce (1983) the mathematical arguments are separated from the horse sense, so that the book may be read with profit by biologists. The subject of experimental design was pioneered and largely developed by R. A. Fisher, and the first few chapters of Fisher (1971) are still an excellent introduction to the philosophy.

8.1. The importance of experimental design

Before the advent of computers, experimental designs were chosen to make the analysis as easy as possible. Using a desk calculator, analysis of a badly designed experiment can take many hours. But a computer can do any analysis, however complex. From that point of view, experimental design is no longer quite so important as it used to be. But it is still very important, for three reasons. First, we naturally want to get as much information as possible for our money. Second, we want to avoid non-orthogonality with its attendant ambiguities (Chapter 2). And thirdly, we must avoid designs which cannot be analysed at all. For example, suppose we wish to try a new drug on some rats. And suppose that we misguidedly give the drug only to female rats, and use only male rats as control. Then the drug-versus-control difference is completely confused, or 'confounded', with the female-versus-male difference, and no amount of statistical analysis can separate the two comparisons. An experiment must therefore be designed to give an unbiased estimate of differences between treatments.[1] Chapter 6 said that to test the significance of a treatment effect, we ought to use a remainder mean square which will indicate how big the difference is expected to be if the treatment has no real effect at all. An experimental design should therefore permit the estimation of an appropriate mean square. In technical language, the design must furnish a valid estimate of residual variance.

Both those purposes—unbiased estimates of treatment effects, and a valid estimate of residual variation—are served by randomization. Some people think that randomization is Fisher's answer to all statistical evils. Sadly, it is not. Randomization only works *on average*—that is to say, a whole series of experiments, each with a different randomization, would on average give unbiased estimates of treatment effects and of residual variance. If we assign each rat to drug or control at random, there can on average be no intrinsic difference between the two sets of rats, because any rat is equally likely to appear in either set. But in practice we are only going to do one experiment involving one particular randomization. If we divide rats at random into two groups of size N, there is bound to be some difference between the two means. According to the rules of Chapter 2, the difference will on average be zero with standard error $\sqrt{(2V/N)}$. Randomization is the best we can do to avoid accidental bias, but if the size N of each group is small, large intrinsic differences between groups may still occur by chance, to bias the estimated treatment effects.[2]

We use randomization to take care of *unpredictable* variation, but the experiment could be designed so that any treatment is applied to equal numbers of male and female rats. It would then be 'balanced' for sexes. We can only do that if we can recognize males and females. Comparisons between treatments would then be free of sexual complications. Alternatively, animals might be assigned to treatments entirely at random, without regard to sex. Each treatment would receive approximately, but not exactly, equal numbers of males and females. Then if there were important differences between the sexes which affected the experimental results, the experiment would be less accurate because the effects of sex would be mixed in with the effects of treatments. Sexes and treatments would not form an orthogonal two-way table (Chapter 2). Once a category—in this case, sex—has been recognized, its possible effects should be allowed for in the analysis (Example 2.5). So the method of analysis follows the design of the experiment (given a statistical model and remainder distribution).

Those, then, are the reasons why the biologist needs to consult the statistician before the experiment is done. Not too many biologists do so, which is perhaps fortunate for the overworked statistician. Very often an experiment could have been done more efficiently. Occasionally it cannot be analysed at all, to answer the questions which the biologist wanted to ask. When a biologist does consult a statistician about an experiment, it sometimes takes an hour to establish exactly what questions are to be asked, and five minutes to design an appropriate experiment. Here we meet the two overriding considerations in the design of experiments and samples. First, if you think out exactly and precisely what questions you wish to ask, an appropriate design is very often obvious. Second, it is wise to consider how the results will be analysed, *before* the experiment is done. There is then less chance of omitting some vital measurement. Those two rules may seem trite until you recognize their overwhelming cogency. It is very sad to find, too late, that the wrong question has been asked or that something essential has not been measured. There is therefore a temptation to measure everything in sight, in the hope that it may come in useful. It may certainly be a good idea to take any extra measurements that can be obtained cheaply. On the other hand, there is no point in measuring things just because we know how to measure them. Such questions are a matter for personal decision, but 'the idea of compiling massive routine records in the hope that they will eventually be of value in retrospective research is nearly always disappointing in practice, though useful suggestions and indications may be obtained' (Bailey 1967). It is common experience that large masses of data, previously collected for some other purpose, rarely afford much information about some new question.

8.2. Design in practice

The discussion so far applies equally to experiments and sample surveys. What is the difference between the two? The difference is causality. When we do an experiment, we ourselves specify which units (e.g. rats) shall receive which treatments, and we try to ensure by randomization that the effects of those treatments are not confounded with any other extraneous effects. We are therefore entitled to assert that the treatments which we applied must have *caused* the observed effects. In a sample survey, on the other hand, we can only observe relations. For example, the density of vole populations is correlated with sodium content of the soil. While we may reasonably use sodium content to predict the numbers of voles, we cannot validly say that the sodium *determines* the vole numbers—both might be caused by some third agent. Only when we ourselves can decide how the treatments are applied may we assert causality. When some Act of God creates an unusual situation, it is not logically valid to regard it as a 'natural experiment' which allows us to deduce cause and effect. But there are two provisos. First, as in the case of smoking and lung cancer, purely circumstantial evidence may be so overwhelming in favour of some reasonable explanation that we should be foolish to deny causality. That argument may not be logical, but it is rational (Chapter 7). Second, causality is a human notion which is unknown to the universe at large. In a complex multivariate case, we may need to consider exactly what is meant by 'causality'. Usually it means that in the past, event A has always entailed event B, and there is some good reason to expect that A will go on entailing B in the future. But in a complex situation, B may follow A only when other conditions are met. This point is especially relevant to ecologists who try to unravel complex natural relationships.

In a sample, as in an experiment, the units are chosen with some degree of randomness. The reason is the same as before—to permit a valid estimate of residual variance. Chapter 3 said that the values of x in a regression may be anything we choose. Therefore, in a sample or experiment intended to estimate a regression, the randomization need only concern the ys. The x-values may be deliberately chosen, either for practical convenience or to improve the accuracy of the estimated regression. This raises the question, how big should the sample be?

Consider the simple case of a comparison of two treatments. If each mean is based on N observations, and the difference of the two means is d, the standard error of d is $\sqrt{(2V/N)}$ and therefore, assuming Normality, $d/\sqrt{(2V/N)} = t$. If we can guess approximately what the residual variance V will be, we can work out what value of N is needed to render 'significant' any value of d which we care to specify. In other words, we can work out the sample size needed on average to detect a treatment difference of any specified size. A similar argument will deduce the sample size needed to estimate a regression to a specified accuracy. A more sophisticated version of this method will work out the larger sample size needed to make 95 per cent certain that a specified difference will be detectable (e.g. Cochran and Cox 1957).

If N turns out to be small, it may be necessary to increase it in order to get a reasonable number of degrees of freedom. Looking at a table of t, you will see that the values are very large when degrees of freedom are few, but that the values of t do not change very much once there are twelve or more degrees of freedom. So the residual mean square should, if possible, have at least 12 degrees of freedom. Occasionally, when dealing with expensive or scarce material—e.g. elephants—we have to make do with very few degrees of freedom, and the experiment or sample suffers accordingly. In any case, we must have some initial idea of the size of the residual variance V. Very often we can guess from previous experience; but if we have no idea how much variation there will be, it is quite impossible to decide in advance how big N should be to give a specified accuracy. It may be necessary to do some preliminary work to assess the amount of variation.

8.3. Sequential methods

Sequential methods offer a way round the difficulty. In a sequential experiment or sample, the work is done in a series of steps, and is discontinued as soon as enough evidence has accumulated about the question that is being asked. This idea is very attractive because it minimizes the amount of work required. There are several reasons why it is not often used in biological research.

First, although the variance V need not be known in advance, the type of remainder distribution must be known. A sequential scheme intended for use on a population with a Poisson distribution will go badly wrong if the distribution is in fact Negative Binomial.

Second, sequential schemes are designed to distinguish between pre-selected hypotheses, i.e. to accept one and reject the others. But Chapter 7 pointed out that research workers sometimes want to assess a situation, rather than jump to conclusions.

Third, it is found in practice that while sequential methods can economically distinguish between categories, sequential methods of estimation save very little work. If the question is 'How many insects are there on this tree?', the sequential and fixed-size samples needed for a specified degree of accuracy will be about the same size. If the question is 'Are there a lot of insects, or only very few?' it may pay to use a sequential sample. It is easy to see why. If there are a lot of insects, only a small effort is needed to establish the fact. In that case the sequential sample will stop short, whereas a fixed-size sample, which has to be large enough to detect 'very few insects', will waste a lot of work. But the overriding objection to sequential samples is the time they take. Consider an experiment on the growth of annual plants. A sequential scheme requires us to grow a few plants this year, a few more next year, and so on until enough have been grown altogether. It is much quicker to grow a large number all at once, even if it turns out to be too many. In biological research most situations are like that. In some cases, e.g. genetics of micro-organisms which have short life-cycles, or surveys of insect abundance, sequential methods can sometimes save a lot of work.

8.4. Recording data

It is worth taking trouble over the way data are recorded. Items written down on odd bits of paper will become unidentifiable or lost. It is very important to label data well, so that there can be no doubt later as to what the numbers represent. Data should be written down in the format which will make them easy to use subsequently. If the observations are going to be extensive, specially designed forms can be very useful; they make the data easily accessible, and immediately reveal omissions. Such forms should leave space for working notes, for corrections, and for extra measurements which you may decide to take. Writing should be in ink, not pencil; mistakes should be crossed out, not overwritten. Sometimes the data can be recorded ready for computer input. Sooner or later, computers will be able to read handwritten characters as a standard method of input. Until then, there is one cardinal principle. If a mechanical method of recording is used, it must give an instantaneous printed record, so that mistakes can be recognized and corrected in good time. There are now portable recorders that will do the job, but they still require considerable self-discipline from the operator to check that the right figures are being recorded. Fully automatic recording devices can be useful, but they can easily produce vast quantities of superfluous data. For example, daily maximum and minimum temperatures are often just as good as continuous temperature records. Once again, the overriding consideration is 'How am I going to use the data?'

Notes

1. Bias, precision, and accuracy. Consider a series of similar experiments or samples, each of which yields an estimate of some population parameter θ. The estimates are unbiased if the average of the estimates—strictly speaking, the average of all possible estimates of that type—equals θ. The estimates are precise if their variation is small, whether or not they are grouped around the true value of θ. And they are accurate if they are both unbiased and precise.

2. The usual type of analysis of experimental data considers those data to be a sample from a whole population of possible values, as in Chapter 1. There is an alternative approach, 'randomization theory', which accepts the observed data as fixed entities and considers instead the population of all possible different randomizations of those same data. If, in the experiment actually performed, large differences appeared between treatment means—and if, in the great majority of re-randomizations of the same data, those differences mostly disappear—it follows that either the particular randomization actually adopted gave very unusual results by chance, or there must be genuine treatment effects.

This argument gets round the problem mentioned in Chapter 4, that the residuals of similar or adjacent experimental units might be correlated—for although the different units might not be independent, the different randomizations certainly are. (But an analysis which recognizes and allows for the correlation between residuals might still be more accurate.) The argument does not get round the basic difficulty that randomization only works on average, whereas any one experiment uses only one particular randomization. The problem is minimized by using several replicates of the experiment, each with its own randomization. There is no final solution to the problem, because it is the problem of inference itself (Chapter 7).

Examples 8

8.1. In Example 2.2 the residual mean square, omitting the erroneous value, was 0.00410. How many pigs per treatment are needed to detect an average difference of (a) 0.05 and (b) 0.01 cm per day between two treatments, using the 5 per cent value of t?

9 Distributions, indices of diversity, and information

9.1. Classifying distributions

During the past forty years, biologists have paid considerable attention to the various types of statistical distribution encountered in the field: for example, the distribution in a sample of insects of the numbers of individuals per species, or perhaps the numbers of species per quadrat in a sample of plants. The analysis was intended to serve two purposes. First it was hoped that, by showing that a given sample conformed to a Negative Binomial, log-Normal, Hypergeometric or Exponential distribution, we might deduce something about the underlying biology which gave rise to that distribution. Second, biologists wanted to measure an 'index of dispersion' which would represent the properties of the distribution, so permitting comparison of one sample with another.

The first of these purposes was over-optimistic. Suppose that some distribution turns out to be Negative Binomial. Such a distribution can arise in many different ways: Southwood (1978) lists five, and there are plenty more. So we cannot deduce how the observed distribution arose in practice. To determine how it did, we have to investigate the biological process itself. The study of distributions does not give quick biological information.

There is also a technical problem. It is very hard to establish that a given sample belongs to some particular distribution, and no other. If the counts were homogeneous, differing from each other only by chance, we should expect them to show a Poisson distribution. The Poisson distribution has only one parameter, the mean m. It is easy to estimate m—we use the average sample count (except when a zero category, e.g. the number of species not represented in a sample, cannot be counted; in that case estimation of m is more complicated). The observed distribution of counts can then be compared with the expected Poisson distribution in two ways. First, the variance of a Poisson distribution is equal to the mean m. So the comparison can be made quickly by comparing the variance of the sample counts with the estimate of m. Alternatively, it can be made more thoroughly by a χ^2 test of goodness-of-fit of frequencies (Example 9.1). If the distribution is not of the Poisson type, it may be under-dispersed, i.e. have a variance less than its mean, implying that some control process acts to reduce the variability.

An example is the number of chiasmata per chromosome in a set of homologous chromosomes. More often, the distribution is over-dispersed, with variance greater than the mean, in which case the counts cannot be homogeneous; different counts are different, not purely by chance, but because of heterogeneity either in the underlying biology or in the method of sampling.

Whereas the Poisson distribution has only one parameter, any under- or over-dispersed distribution must have at least two parameters, to give the right values for mean and variance. A theoretical distribution with two adjustable parameters can be made to fit a much wider range of observed distributions than can the Poisson. Even so, some sets of data show a relationship between variance and mean (cf. Note 2, Chapter 4) which is not matched by any standard theoretical distribution. In practice it is often easy to show that a given sample does not obey a Poisson distribution (one parameter), but difficult to prove that it is not, say, Negative Binomial (two parameters). It may take a sample of several hundred counts to distinguish between a Negative Binomial and a Logarithmic distribution. And when we have such a large number of counts, they often do not conform to any of the standard distributions. So it is difficult in practice to identify a particular distribution with certainty. Nor is there much point in doing so. It is useful to show that the observed distribution is not Poisson, since that implies some kind of non-randomness. What causes the non-randomness is then a biological, not a statistical, problem. It is useless to fit, say, a Negative Binomial distribution unless the parameters of that distribution are to be subjected to further analysis. The Negative Binomial is then used simply as shorthand to sum up the observed set of counts. It would not much matter if the true distribution were not exactly Negative Binomial, provided that it cannot be distinguished from one in practice.

There is an amusing special case. If a series of taxonomists, some 'splitters' and some 'lumpers', worked in succession on a family whose members actually showed purely random variation, the resultant distribution of numbers of species per genus would be Logarithmic. That is precisely the distribution actually found in taxonomy. In justice to taxonomists, we must remember that a given distribution can arise in more than one way.

9.2. Indices of diversity

Although we no longer expect to use statistical distributions to get easy biological information, we still want to compare one distribution with another. If the distributions are Poisson, we can compare their means. But if a distribution is over-dispersed—and in biology it usually is—it takes a mean and a variance to describe it, and possibly a measure of skewness too. An 'index of diversity' is intended to summarize, in a single measure, the degree of dispersion of the population. The index can then be used to characterize and compare different populations. It is referred to as an index of dispersion when it considers the distribution of one species, and an index of diversity when it refers to the composition of a whole flora.

In general, such indices attempt the impossible, namely to summarize in a single parameter the complete distribution which specifies the frequencies of all possible counts, from zero to infinity. Obviously, any one index can represent only one particular aspect of the distribution, and different indices will measure different aspects. An obvious candidate is the sample variance, or perhaps the coefficient of variation (Note 2, Chapter 4). But suppose we are dealing with a set of sample quadrats of a plant population. We want an index which will characterize not just the sample, but the whole population. In other words, the index must be unaffected by the number or physical size or shape of the quadrats. That is very difficult. Before we can be sure that a given index is unaffected by the sampling procedure, we need to know the distribution, not just of the sample values, but of the original population—but usually all we know of the population comes from the sample itself. Supposing we have calculated an index of diversity, what use is it? It can be used only to compare two different populations, and that can only be done safely when the two populations have similar statistical distributions and are sampled in the same way. So the idea of an absolute index of dispersion is usually over-optimistic. Greig-Smith (1983) discusses indices of dispersion and diversity at length.

9.3. Information

One possible index is the measure of 'information'. Suppose that the individual units of a sample are sorted into categories. For example, individual animals might be sorted into their various species. If the frequency of the ith category is p_i, the 'information' is $-\sum p_i \log_2(p_i)$. This is the measure of information used in communications engineering (it is different from, although related to, Fisher's definition of information in statistical theory). This measure of information is very closely related to entropy in statistical mechanics, and to likelihood in statistical theory. But there is an important practical difference. In communications theory, the information measures how much information we ourselves have to supply on average, to specify a given message; and similarly in statistical mechanics, the entropy measures our own ignorance of the

energies of individual molecules: but in statistics the likelihood sums up
the information which the sample gives to us.

The information $-\sum p_i \log_2(p_i)$ is the average number of choices, or
guesses, needed to allocate an individual to its category. Suppose there
are three categories A, B and C with frequencies $\frac{1}{2}$, $\frac{1}{4}$ and $\frac{1}{4}$. Then

$$ - [\tfrac{1}{2} \log_2(\tfrac{1}{2}) + \tfrac{1}{4} \log_2(\tfrac{1}{4}) + \tfrac{1}{4} \log_2(\tfrac{1}{4})] $$

is $\frac{3}{2}$. We wish, by a series of guesses, to allocate a new individual to its
proper category. The argument is restricted to binary yes-or-no choices
because communications engineering works that way: dot or dash,
positive or negative, presence or absence. We first ask 'Is it A?' If the
answer is yes, we have found the right category with one guess. The
probability of that happening is $\frac{1}{2}$. If the answer is no, we must ask a
further question 'Is it B?' The answer will decide between B and C since
we know beforehand that there are only three categories. So it takes two
questions to identify B or C, and the extra question will be needed on
average in $\frac{1}{2}$ of all cases. The average number of questions necessary is
$\frac{1}{2} \times 1 + \frac{1}{2} \times 2 = \frac{3}{2}$, which is the information required on average to
distinguish A, B and C. It tells us how much information we ourselves
have to supply to specify an individual's category. It does not tell us how
much information a sample contains about some question which we
should like to ask. It is a long-term average, since in no case can we have
exactly $\frac{3}{2}$ guesses.

It is quite another matter when the same function $-\sum p_i \log(p_i)$ is used
as an index of diversity. Taking a sample of N individuals, we allocate
each to a category, with n_i in the ith category. Then $p_i = n_i/N$. The more
categories there are, and the more evenly the individuals are scattered
among the categories, the greater is the value of $-\sum p_i \log(p_i)$. So that
value is indeed one measure of the diversity of the sample. But the value
is no longer a long-term average, and no longer has a theoretical
interpretation as an absolute measure of our uncertainty about the
identity of an individual. The values of p_i are estimates from a finite
sample. If we increased the sample size N, we should expect to encounter
new categories not represented in the existing sample. The estimated
value $-\sum p_i \log(p_i)$ may or may not be a good measure of the diversity of
the entire population. There is no longer any theoretical backing to show
that it is an absolute measure. By analogy it is an attractive candidate as
an index of diversity, but it must be treated in practice with the same
caution as any other index.

Until recently, plant ecology has been largely concerned with patterns
of vegetation, and indices of dispersion and of diversity have provided a
happy hunting ground for plant ecologists. The emphasis in plant
ecology, as in animal ecology, is now on population dynamics
(Chapter 11), and so there is less emphasis on indices of diversity. It is

still important to realize that any such index depends, as so often in mathematical biology, on assumptions which are rarely stated explicitly.

Examples 9

9.1. The number of bacterial colonies was counted on each square centimetre of an agar medium.

Number of colonies (y)	0	1	2	3	4	5	6	7	8
Observed frequency	59	86	49	30	16	2	0	1	1

The sample contains 59 values $y = 0$, 86 values $y = 1$ and so on.
(a) Show that the sample mean of y is 1.488 and the mean square 1.773. If the distribution is Poisson, the variance equals the mean. Test the variance ratio 1.773/1.488—how many degrees of freedom?
(b) The probability of counting r colonies when the mean is m, is $e^{-m}m^r/r!$ if the distribution is Poisson. Show that the expected frequencies are

0	1	2	3	4	5 or more
55.12	82.00	60.99	30.25	11.25	4.39

Compare the observed and expected frequencies by χ^2. The value of χ^2 will have four degrees of freedom—why?

Methods (a) and (b) both test for a Poisson distribution. Method (b) takes longer than (a), but it is more thorough because it checks the whole distribution, not just the variance. But it is unlikely that a distribution of counts should differ seriously from the Poisson if its variance equals its mean. So the variance ratio is usually quite good enough to test for a Poisson distribution.

10   Quantitative genetics

This chapter and the next consider two fields of biological research which rely heavily on statistical analysis. The two topics each have intrinsic interest, but more generally they show the limitations of biometrical analyses, including the danger of believing your own assumptions.

10.1. The basic model

Quantitative genetics studies the inheritance of characters such as individual size or weight, which show a continuous range of variation. Each y-value is influenced by the combined effect of many different genes whose individual contributions are not distinguishable. Unlike the case of Mendelian inheritance, it is not possible to determine the genotype by looking at the phenotype. The object is to predict, as accurately as possible, from one generation to the next: so the basic question is 'If I am so tall and my spouse is so tall, how tall will our children be?' The answer must depend partly on genetic and partly on environmental effects. Therefore, the simplest possible model is

phenotypic value of y = a genetic effect + an environmental effect.   (10.1)

We are free to choose a scale of measurement of y, if one can be found (Chapter 5), which will guarantee the additivity of the two effects in eqn (10.1).

Now consider a one-way table in which the different treatment categories are different genotypes. In other words, there is a set of different genotypes, each represented by a clone of several individuals. From eqn (10.1) it follows that

> the treatments mean square is expected to equal a genetic
> component of variance + an environmental component
> of variance,   (10.2)
> the residual mean square is expected to equal the
> environmental component only.

This is a simple example of components of variance (Note 5, Chapter 3). By subtracting the residual mean square from the treatments mean square, we get an estimate of the genetic component of variance. If, as occasionally happens, the treatments mean square is actually less than the residual mean square, the genetic component of variance comes out negative, which it shouldn't. That can happen either because of the inaccuracy of mean squares as measurements of variance (Chapter 4) or because the two effects in eqn (10.1) interact. The genetic component, expressed as a fraction of the genetic + environmental components, is called the heritability, because it indicates how much of the variance of individuals is heritable. For technical reasons, the analysis is slightly more complicated than the above, but the principle is the same.

The estimate of heritability relies heavily on the truth of eqn (10.1). That equation asserts that if two genotypes are grown side-by-side in a series of different environments, the difference in performance of those two genotypes will always be the same. So the scale of measurement of y is chosen to minimize the interactions between the genetic and environmental effects. It is not always possible to find a scale of measurement on which there are no interactions at all. The crucial assumption in eqn (10.1), that the two effects add together, may not be justified on any scale.

The object of the analysis, as usual, is to make predictions. In this case, the desired predictions are of two kinds: to predict values for children from those of their parents, and to predict responses to selection which is to be applied for several consecutive generations. The two kinds of prediction, if they are possible at all, require rather different methods. The first can be done purely statistically: the second inevitably requires some understanding of the genetics. Either method involves further analysis of the genetic effect in eqn (10.1).

10.2. The midparental genetic pattern

Underlying the whole subject is the 'midparental' pattern of inheritance, discovered by Francis Galton in 1886. Consider any quantitative character y, split into a genetic value plus an environmental value (eqn 10.1) for each individual. It is observed in practice that for any pair of parents, the genetic values of their progeny are on average about half-way between the genetic values of the two parents. This empirical midparental genetic pattern arises because in most sexual species, the genetic contributions of male and female to each progeny are roughly equal. The midparental pattern is known to racehorse breeders as 'Galton's rule'. In a few species, such as dog roses, the genetic contributions are very unequal and the midparental pattern does not hold.

The midparental pattern refers only to the mean of all the progeny in one family. Of course the individual progeny, far from being identical, are very variable. The midparental pattern is found to be approximately correct, on any reasonable scale of measurement. Suppose for the moment that there exists a scale on which it is exactly true. And let the genetic variance of y among all individuals in the population be V. Then the genetic variance of the midparental values is $V/2$, because the midparent is the mean of two individual values. This assumes no correlation between the parental values, or that mating is random with respect to y.

It follows that if the population variance is to remain constant from generation to generation (as it does), the genetic variance of the individuals within any one family must also be $V/2$, so that the

total genetic variance V = between-families $V/2$ + within-families $V/2$.

This means that, given the midparental pattern of inheritance, the population genetic variance will be twice the genetic variance of brothers and sisters. So the within-family variance, which arises from sexual segregation, determines the genetic variance of the whole population. The argument derives solely from the midparental pattern: it makes no assumptions about the effects on y of individual genes. In practice it need not be exactly correct, either because the midparental pattern is not exactly observed, or because of non-random mating, or because natural selection, favouring some individuals at the expense of others, curtails the variation of the population. Fisher (1930) states the argument as follows: 'of the heritable variance in any character in each generation a proportion is due to the hereditary differences in their parents, while the remainder, including nearly all the differences between whole brothers and sisters, is due to segregation. These portions are not very unequal.'

The midparental pattern applies to the genetic values in eqn (10.1), not necessarily to the phenotypic values y. If the parents have been chosen because they have unusually large or small values of y, their outstanding size will be partly genetic, and partly non-genetic or 'environmental'. Only the genetic value is transmitted to the offspring, and therefore only that fraction of the parent's outstanding size y is reflected in the progeny's average y. In so far as a parent is chosen because it has a large non-genetic contribution to y, its progeny will show no corresponding deviation from the population mean. Therefore, in crosses between parents specially chosen for their outstanding phenotypes, the values of y for the progeny are generally less extreme: they 'regress' towards the population mean. Galton first used the term 'regression' in this context: the slope of the regression should equal the heritability, provided that eqn (10.1), and the midparental pattern, are both correct. (In the extreme case where there is no genetic variation, i.e. the heritability is zero, there will be no correlation at all between the phenotypic values of parents and offspring.) Thus the slope of the regression of progeny on midparent, calculated from phenotypic values of y, is not 1, but is diluted by environmental variation: just as errors of measurement in x dilute the regression of y on x as a measure of the 'functional relation' between Y and X (Chapter 3). The term 'regression', first introduced in this genetical context, is now used very generally in statistical practice.

The midparental pattern forms the basis for predictions from parents to offspring. Since there is so much variation within one family, predictions of individual children cannot be accurate. The best that can be done is to predict family means, and recognize that the difference between any one individual and its family mean is wholly unpredictable. The whole apparatus of sexuality, with its built-in randomization, ensures that!

The first type of prediction is by simple regression. Pairs of male and female parents, with known values of y, are mated to produce families of progeny whose values of y are also measured. The midparental phenotypic value, i.e. the mean of the two parental values of y, is calculated for each pair; and the regression of family means on midparental values can be used to predict the results of further crosses which have not, in fact, been made. This is a straightforward application of regression which is based on the midparental pattern—diluted because we are measuring phenotypic, not genetic, values.

Another method of prediction, by 'combining abilities', is slightly more complicated. A series of crosses is made between a set of parents, or between a set of parental lines each containing several individuals with closely similar genotypes. The situation is analogous to a two-way table with the male parents as rows and the females as columns, except that the same parents or parental lines may appear as both males and females. The analysis (Yates 1947) consists of fitting additive combining abilities, equivalent to the row and column constants of a two-way table. The analysis does not include measurements of the parents themselves, but only of their progenies. The 'combining abilities' estimate the genetic values of each parent, as transmitted to their progeny.

10.3. The additive genetic model

The two types of analysis described in the previous two paragraphs, are purely statistical in nature. They are robust because they are concerned with means and regressions, not variances. They make no genetical assumptions beyond the midparental pattern of inheritance, which can itself be checked statistically. Beyond that, they tell us nothing about how the underlying genetics works. Any more penetrating analysis must interpret the variation in terms of the effects of individual genes. This is where the trouble starts. Equation (10.1) is extended by assuming that the genetic effect is itself the additive sum of the independent effects of all the individual genes which affect y. Anyone who has watched a baby grow will object that that is a preposterous assumption. Development and growth is a highly coordinated process involving close interaction between vast numbers of genes. The genes plainly do not act independently. But here we are concerned, not with the process itself, but only with the statistical effects of the genes on the final product. All the same, there is every reason to expect that interactions between gene effects will be as important as the effects themselves. The theory was worked out by Fisher (1918) who explicitly admits that unless most of the interactions can be neglected, further analysis is impossible.

We can now see why, in the past, quantitative genetics has relied heavily on the analysis of variances *per se*, despite the statistical objection that estimates of variance are neither accurate nor robust (Chapter 4). Compare the case of a lot of genes each with small effect with that of a few genes of large effect. The mean of y will be the same in both cases, but the variance will differ. So we are forced to use variances *per se*, if we want information about individual gene effects.

The offspring/midparent regression described above can predict the response to selection, at least for one generation. If we choose parents with the largest available values of y, the regression will predict how big their progenies' values will be, on average. This is a purely statistical prediction, perfectly reliable provided that the regression is correctly estimated. But estimation of the statistical regression takes time and work, and it predicts only one generation ahead. Is it not possible to make use of our knowledge of genetics instead? Assuming that gene effects are additive, and assuming further that mating is random—i.e. that different genes are assorted randomly—the heritability, as estimated from components of variance earlier in the chapter, may itself be used to predict how much a population will respond to selection pressures. Falconer (1981), the standard textbook on the subject, gives the details. In fact, on those assumptions, the slope of the offspring/midparent regression should equal the heritability. Usually it is rather less.

Assuming that gene effects are additive, and given the heritabilities, the theory can predict the correlations between values of y measured on relatives of any specified degree: brothers and sisters, first and second cousins, grandparents and grandchildren (Fisher 1918).[1] It is very commonly found that such predictions are perhaps 80 per cent correct, but not perfect. So the effects of genes cannot be perfectly additive, but they may be nearly so. The genetic component of variance in eqn (10.2) is therefore sub-divided into 'additive' and 'non-additive' components which together can account almost perfectly for the observed correlations between relatives. The 'additive' component takes the lion's share. It is commonly, but quite fallaciously, assumed that the 'additive' component must indeed arise from additive effects of genes. This is not so, for the following reason.

10.3.1. Statistical inseparability of the additive genetic model and the midparental pattern

The midparental pattern itself predicts what the genetic correlations between relatives should be. For example, the genetic value of any progeny individual will be

$$\text{midparental value} + \text{deviation},$$

where 'deviation' is the unpredictable genetic deviation from the family mean, due to genetic segregation. Any two full siblings (brothers or sisters) will share the same midparental value, but will have unrelated 'deviations'. But, as argued above, the variance of the 'midparental value' is $V/2$. It follows that the genetic correlation between full siblings is $\frac{1}{2}$, if the midparental pattern is exact. Similarly, the genetic correlation between one parent and its progeny is also $\frac{1}{2}$; between half-siblings, or between grandparent and grandchild $\frac{1}{4}$; and so on. These figures refer to the correlations between the genetic values in eqn (10.1), not to the phenotypic correlations.

Now suppose that genes do have strictly additive effects, without interactions of any kind. Then the predicted genetic correlations between relatives are identical with those predicted by the midparental pattern, i.e. $\frac{1}{2}$, $\frac{1}{4}$ etc. (Falconer 1981). Therefore, as long as we are analysing correlations (or equivalently, covariances), it is impossible to distinguish between (a) the theory of genes with additive effects and (b) the simple midparental statistical pattern. It follows that the theory of additive gene effects is wholly superfluous, since the statistical pattern, which makes no unproven assumptions, gives identical results. (It is true that the additive gene theory is usually extended to include one particular kind of interaction, viz. dominance, which the midparental pattern does not recognize, but predictions from parents to offspring are not thereby improved.) So it is not possible, in practice, to distinguish the theory of additive gene effects from the simple midparental statistical pattern.

It may be objected that the success of additive-gene-effect models in predicting immediate responses to selection (Falconer 1981) confirms that the effects of individual genes must be approximately additive. Not so. The correlations between relatives, mentioned above, ensure that the regressions used to predict selective response will be near enough correct, given only the midparental pattern. Predictions of selective advance depend on that pattern, and not on additivity of gene effects.

Alternatively, it may be argued that the midparental pattern is insufficient to explain the special analyses of crosses which derive from inbred stocks (Mather and Jinks 1982). For example, the genetic variance of backcrosses is generally about half that of F_2, an observation which the midparental pattern, on its own, cannot explain. But such observations merely confirm that the underlying genetics are of the diploid chromosomal type. They tell us nothing about individual gene effects, unless we assume, for no good reason, that most interactions can be ignored.

If gene effects *were* truly additive, the midparental pattern would necessarily be exactly correct. (The converse is not true: a perfect midparental pattern does not necessarily imply additivity of gene effects.) In that case the empirical data, when analysed in the usual way (Falconer 1981), would be judged perfectly 'additive', with no 'non-additive' component at all. Thus 'additivity', in practical quantitative genetics, measures the extent to which family means adhere to the midparental pattern: and 'non-additivity' measures the departures from that pattern. 'Non-additivity' is simply the extent to which the data violate the midparental statistical pattern, and nothing more. It says nothing about the additivity, or otherwise, of gene effects.

This sad story underlines a very general point about statistics. Even though statistical analyses may describe the data well, they need not represent the underlying biological processes. For example, the analysis of fertilizer trials, in terms of additive effects, tells us nothing about the complicated physical pathway from a dose of fertilizer to its effect on crop yield; and in the present context, the success of 'additive' analyses tells us nothing in detail about the underlying genetics. To learn about biological processes, we have to study those processes directly.

It may appear, then, that Galton's midparental pattern still comprises all that we certainly know about quantitative genetics. That is largely true, but in the past 30 years a further feature has come to light: the amount of variation is not free-for-all, but is itself under genetic control, and is restricted within fixed patterns (Gilbert 1988). In this respect, quantitative genetic variation is more complicated than simple Mendelian variation. Fisher's (1918) theory, which is based on binomial frequencies of individual genes with fixed effects (Falconer 1981, Mather and Jinks 1982), does not recognize that complication. Any realistic theory of quantitative inheritance must now recognize that there are two levels of genetic variation. There is the usual variation between individuals, and superimposed on that, there is genetic control of the amount of that variation. There are consequently two levels of selection too.

The practical applications of quantitative genetics lie in the fields of plant and animal breeding, and of population biology. Some breeders used to hope that quantitative analyses could furnish information about gene effects which would reveal the best plan of breeding work. That hope has proved vain, but the purely statistical methods of offspring/parent regression and combining abilities have sometimes proved useful, although usually they merely confirm the old adage 'breed from the best'. It does not pay a breeder to examine a genetical situation in detail, if he has to wait a long time for the answers. But some kind of analysis is essential for characters which cannot be measured directly, e.g. the merit of a bull for milk production. It is still widely, but erroneously, believed that 'additive genetic variance' does measure additive gene effects, despite Falconer's (1981) explicit disavowal of that proposition. Meanwhile population biologists are devoting more and more attention to 'life history traits' which directly affect reproduction and survival. Since such traits are invariably quantitative, the methods of quantitative genetics are needed for their analysis (Chapter 11). For this purpose we do not need to understand the underlying genetics, but only to predict from parents to offspring, so the statistical methods of this chapter are perfectly adequate.

Note

1. Fisher (1918) analysed the correlations between human relatives for three physical measurements, using the model of additive gene effects with complete dominance and 100 per cent heritability (no environmental effects). However, those same correlations are equally well interpreted by the midparental model (no dominance) with $75(\pm 2.5$ S.E.) per cent heritability, and a correlation of $0.3(\pm 0.07)$ between the environmental effects on brothers and sisters, who are reared in a similar environment. These two different scenarios fit the data equally well. This underlines the point that a statistical model, however consistent with a set of data, need not describe the underlying biology.

Examples 10

10.1. Equation (10.2) divides the treatments mean square into a genetic component of variance + an environmental component of variance. The estimate of the genetic component may, with bad luck, come out negative. On the other hand, in an analysis of variance the total sum of squares is divided into a treatments sum of squares + a residual sum of squares, neither of which can possibly be negative. Explain the difference.

11 Population biology

Population biology is an enormous subject. This chapter will consider some statistical aspects—especially the assumptions involved. The first assumption, widely accepted by ecologists, is that the subject is concerned with the 'distribution and abundance of animals and plants'. Until recently, animal ecologists have concentrated on *abundance*, i.e. on population dynamics, and have generally paid less attention to geographic distribution—except on the largest scale, when it becomes the province of zoogeographers. And until recently, plant ecologists have done the opposite, concentrating on *distribution*, i.e. on associations either with the features of geography or with other plants (Greig-Smith 1983); but now there is much more interest in plant population dynamics (Harper 1977). In this chapter the word 'population' refers to a lot of real live animals or plants, not just a statistical abstraction.

11.1. Theoretical ecology

The most remarkable assumption in quantitative ecology underlies 'theoretical ecology', a body of theory based (in current practice) on the proposition that we can deduce how animal and plant populations work by making up mathematical equations in our heads, without reference to empirical data! Such attempts began with the early work of Lotka[1] and Volterra, and the past 20 years have seen much theorizing of this kind, following the comprehensive lead of Goel *et al.* (1971). Since these theories have no empirical foundation, their value depends entirely on the assumptions embodied in the mathematical equations. For example, there has been much discussion of the stability of population dynamics. Here 'stability' means a tendency for population size to return to a steady-state equilibrium. It is easy to make up mathematical equations which predict the existence of an equilibrium, and then to deduce the conditions necessary for such stability; but there is little or no reason to suppose that natural populations actually possess a possible steady-state equilibrium, let alone that it is stable (Connell and Sousa 1983). Population ecology certainly needs, and at present does not have, a coherent theory, but that theory must be intimately based, as in physics and chemistry, on empirical reality.

11.2. Data collection

Any discussion of population dynamics evidently requires successive estimates or indicators of population size. The earliest studies consisted of little more than a series of such estimates. It is now understood that population dynamics cannot be described in terms of total numbers alone—it requires knowledge of age distributions and of the corresponding schedules of reproduction and survival. All these things have to be estimated, preferably in the field, if we want to make sense of the dynamics.

It is rarely possible to make complete counts of animal populations. Many animals hide in the local vegetation or in the water, and they don't stay put. Counts of plants are possible, provided that each plant can be distinguished from its neighbour. Otherwise there are many possible sampling methods, each suited to particular kinds of animal and plant (Southwood 1978). Direct sampling, i.e. the harvesting of all animals and plants on randomly chosen sample plots, is perfectly feasible for plants but may require special drop-traps to prevent the escape of animals. Direct sampling is generally too laborious when the species under investigation is rare. All the indirect methods of estimating population size depend on various assumptions. Some kinds of trap, such as aerial suction traps, give the animals no choice. Ordinary trapping methods, where the animals themselves choose whether or not to enter the traps, will give a true picture of the population only if all animals are equally likely to enter the traps—but it is found that different types of trap, used simultaneously on the same population, give different compositions of age, sex etc. Such trapping methods can be (although they rarely are) calibrated against populations of known size and structure, but there are endless difficulties due to trap-shyness (disinclination, or alternatively over-eagerness, of previously trapped animals to re-enter a trap), marking of traps by animal odours, and unknown trap area, i.e. the area of land from which the animals might enter a given trap.

Capture–recapture methods involve several assumptions (Southwood 1978), of which some can be tested and others can not. Capture–recapture gives very inaccurate, and often biased, estimates of population size unless nearly all the animals in the population are captured and marked, i.e. unless the method approaches complete enumeration. But estimates of survival rates, obtained by following the fortunes of a group of marked animals, are more accurate. Even then, if the animals are very mobile, it is hard to distinguish death from emigration.

11.3. Data analysis

Until recently, the study of population dynamics involved the collection of a series of population estimates, which were then subjected to different kinds of statistical analysis. If the estimates included age distributions, life-tables might be constructed showing the reproductive and survival rates of each age group (Caughley 1977). If the estimates were of total population size only, they were subjected to 'key-factor analysis' which in its various forms is a multiple regression intended to identify the 'key-factors' most important in determining the dynamics of the population. The expression 'key-factor' may mean either an environmental attribute (including population density itself, since the population is part of the environment of each individual) or a particular period in the life-cycle. Such analyses encounter technical problems due to non-independence of the data (Southwood 1978); but the biggest objection to them is that they assume that 'key factors' do exist, which is equivalent to assuming additivity in the multiple regression equation (Chapter 4). Recent studies of the detailed dynamics of various species have shown that there are no 'key factors'; rather, the various influences on the dynamics interact in complex ways. This underlines a point made in previous chapters: simple statistical analyses may describe the data quite well, without conveying any understanding of the underlying biological processes.

During the past twenty years, the rise of the digital computer has permitted a more penetrating approach to the analysis of population dynamics: simulation models of the ecological processes (development and growth, reproduction, predation etc.) which affect numbers. This method requires much more intensive field work than before, and in return it gives much better understanding of the dynamics. Many such models have been attempted, and it is now clear that it is perfectly possible to reconstruct the dynamics of local populations. Reconstruction of *movement* is more difficult. In any case, the hard work lies in the experimental work needed to unravel the ecological complexity, not in making the model itself. The fact that such models can indeed be made shows that population ecology is a technically feasible subject—it is not hopelessly complex and difficult. It is often argued that such models will tell us how to manage plant and animal populations for pest control and wildlife harvesting. If that were so, models would by now be an important tool in practical management, which they are not. Experience so far is that in practice, detailed ecological understanding does not assist management. The essential knowledge, it appears, is of the natural history—where do I live and what do I eat? The chief utility of simulation models has been to assure us that our understanding is reasonably complete, so the paradoxical situation arises that your model is most useful when it gives the wrong answer, indicating that something un-recognized must be going on. Of course, simulations are now used not just for population dynamics, but in other areas of biology too, such as the growth of crop plants.

11.4. Why theoretical ecology fails

These models have made one thing very clear. It takes 1–3 dozen parameters to describe the dynamics of one local population. In technical terms, the dynamics occupy a parameter space of 1–3 dozen dimensions. Gilbert (1982) listed 12–14 parameters for one particularly simple case. By contrast, the theories of physics involve only a few variables or parameters: $PV = RT$ has three variables and one constant, while $E = mc^2$ has two variables and one constant. Consequently, there is a simple reason for the failure of theoretical ecology, whether of the mathematical or non-mathematical kind (McIntosh 1985): it oversimpli-fies. It tries to get a gallon into a pint pot, by reducing the parameter space of several dozen dimensions to one of three or four. For example, there is no known live population whose dynamics can be described in terms of total population size, without sub-division into age groups, sexes and morphs—yet theoreticians regularly make up equations of the type $dN/dt = f(N)$, which assume that the rate of population increase is a function of total population size N. Similarly, the vague classification into 'r-selected' and 'K-selected' lumps together many different aspects of population biology. Ecology must be the only branch of science in which theoreticians do not feel obliged to consider whether their theories correspond to reality!

The *coup de grace* was administered by Wang and Gutierrez (1980), who showed that conclusions about the 'stability' of predator–prey relationships (assuming that they *are* stable in the usual sense—see above) depend critically on the type and extent of environmental fluctuation. Consequently, no realistic general theory of predator–prey dynamics is possible: each case must be considered against its appro-priate environmental background. It will not surprise biologists to learn that predators and prey are adapted to their environment, including the regular fluctuations in that environment.

The overall conclusion, therefore, is that population dynamics is innately complex, and requires a parameter space of at least a dozen dimensions. We have no satisfactory mathematical tools for handling that degree of complexity, except in strictly linear cases. This is the basic reason why valid generalizations or principles have proved so elusive in population ecology (McIntosh 1985). Ecologists can take some comfort from the fact that population dynamics involves only 1–3 dozen, and not 100–300, dimensions. If the latter, ecological analysis, even of a single population, would be impossible.

11.5. Recent studies

During the past ten years, a dozen studies have compared the dynamics of several different populations of the same species. The species involved include birds, mammals, amphibians and insects. In every case without exception, each local population has its own characteristic dynamics, radically different from the dynamics of other populations. So there is no such thing as 'the' dynamics of a species. In one case (Ehrlich and Murphy 1981) it is not just the numbers but the ecological mechanisms which vary from population to population. In another (Gilbert 1982) the animals can adapt to the local food supply to produce densities which vary from population to population by an order of magnitude, yet in every case the population is almost perfectly adapted to the local conditions. So the animals are very well adapted to a range of environments, but the dynamics in each local environment are opportunistic and unpredictable. There are two conclusions. First, to study the dynamics of one species, we have to investigate a dozen different populations (to get the variation in space) for a dozen seasons or generations, whichever is the longer (to get the variation in time). The requirement for a dozen populations and a dozen seasons arises because one dozen provides a reasonable estimate of variation with a reasonable number of degrees of freedom (Chapter 8). Such an investigation is a formidable task, which has rarely been attempted, yet it gives a picture of one species only. The second conclusion is that perhaps the first assumption in this chapter—that population biology should study distribution and abundance—is wrong. If the animals and plants are concerned, not with numbers *per se* but with local adaptation, should not ecologists take the same line?

A similar conclusion may be reached in a different way. There is growing interest in 'life history strategies': why do salmon breed only once in their lives, and humans (in most societies) all the time? When winter comes, is it better to emigrate or enter diapause? The subject began with purely theoretical arguments devoid of empirical backing, but now there is growing interest in the practical study of the 'life history traits' directly concerned in survival and reproduction. Such study involves a mixture of quantitative genetics and population ecology. The prospects for this new approach are quite good, but already it has encountered the same trouble as before. The various traits are inter-connected (Dingle and Hegmann 1982), and must therefore be studied all together. Consequently we are back in a parameter space of a dozen or more dimensions, i.e. the innate complexity of population ecology is just as baffling as ever. We cannot expect to identify valid generalizations or principles (even if they exist) unless some way can be found of orthogonalizing the parameter space, i.e. of selecting out some combinations of the life-history traits that can be considered independently of the rest.

Fortunately, in insects at least, certain parameters can themselves be considered independently, because they affect the dynamics at different times of year (Gilbert 1988). Those parameters are found to obey general rules, so that valid generalizations are indeed possible in population ecology. Rutherford's dictum that 'all science is either physics or stamp-collecting' is not, after all, entirely true of ecology. But in insects at least, the generalizations refer, not to the dynamics as such, but to the underlying life-history parameters. Then the dynamics are determined by the interaction between those parameters, and the environmental variation. Although different local populations may have the same parameter values, they have different dynamics because they inhabit different environments. In the next twenty years we may expect to see a switch of emphasis from population numbers to life history adaptations. This new approach, which incidentally requires just as much statistics as before, should largely replace—and effectively merge—existing approaches to population genetics and population ecology. It is perhaps ironic that plant ecology has switched from distributions to numbers, just when the importance of numbers as such in animal ecology is receding.

11.6. Stochastic processes

The distinction between deterministic and stochastic theory applies throughout biology, and is particularly relevant to population biology. If ducks lay on average 4.6 eggs each, the deterministic theory says that each individual duck lays exactly 4.6 eggs. Stochastic theory takes the more reasonable view that different individuals may lay 1, 2, 3, 4, 5, ... eggs. The stochastic theory therefore requires more information than the deterministic. It has to know not just the average number of eggs, but the distribution of eggs per individual. A duck which lays six eggs is likely to have many more grandchildren than a duck which lays only one egg. So in the stochastic theory of populations, the variability of the different possible outcomes builds up from generation to generation. In other words, starting from a given situation, predictions of the outcome become more and more uncertain as time goes on. At the price of requiring the extra information, a stochastic theory will predict the variability to be expected after several generations, whereas the deterministic theory can only predict an average population size.

We may here recall a very important point in Chapter 1. It is tempting to regard stochastic variation as mere error—indeed, some theoreticians have even treated biological variation as 'white noise'. In fact, the variation is a genuine part of the biology. It can be included in a stochastic model only when its characteristics have been studied empirically.

If the various biological processes which determine the numbers of animals are inter-related linearly, the deterministic answer must equal the average stochastic answer. But in biology relations are not often linear, and so it is theoretically possible that the deterministic answer will be far removed from the true average. In fact, animal numbers generally fluctuate much less than might be expected from those animals' observed powers of increase. Natural stochastic variations are generally small, and so we may reasonably expect the deterministic answer to be fairly representative in most real cases (Southwood 1978). But this conclusion may be too complacent. Just as in ordinary statistical analysis (Chapter 1), stochastic theory divides biological processes into a deterministic skeleton (F_i) with superimposed stochastic variation (residuals). But the residuals, although regarded as random, really represent genuine biological variation which is unpredictable only because we don't understand in detail how it arises. There has to be some physiological reason why one duck lays four eggs and another five. And nearly all studies of population dynamics have dealt with fairly large populations, mainly because of the technical problems of sampling sparse populations. But in large populations the deterministic element naturally predominates: such populations are relatively immune to the slings and arrows of outrageous fortune. So the conclusion that stochastic variation is relatively unimportant may merely reflect this inherent bias in the available evidence in favour of large populations. It is certainly true (Ehrlich and Murphy 1981) that extraordinary historical events can determine the present-day distribution of a species, not just on the zoogeographic scale (e.g. the dominance of marsupials in Australia) but on the ecological scale too.

Stochastic theory can provide insight which the deterministic theory completely misses. Consider a closed animal population with equal birth and death rates. (*Every* surviving population must, in the long run, have approximately equal birth and death rates.) The deterministic theory says that the population numbers will remain constant because births exactly balance deaths. The stochastic theory says that the probability of extinction will steadily increase as time goes on, so that any such population can be expected to go extinct at some finite time in the future—there is no chance that it will survive *ad infinitum*.

It is easy to see why. At every generation there must be some probability, even if small, that the population vanishes, and once that has happened the population cannot re-establish itself. So extinction is an inescapable trap which always lies in wait. If we started with a whole set of such populations, we should find that after some time a lot had gone extinct, but a few had waxed and multiplied. The stochastic average would agree with the deterministic steady-state, but the stochastic distribution of possible outcomes would be very skew in favour of zero. Extinction is inevitable even when the birth and death rates are density dependent. It is a fallacy to suppose that density dependence can save a population from extinction in the long run. But with density-dependent survival, a population will take very much longer on average to go extinct than the same population with density-independent survival. Therefore, stochastic theory tells us that the crux of the ancient argument about density dependence is 'How long do real live populations take on average to go extinct?'—which, of course, we don't know.

If stochastic analyses can do better than deterministic ones, and never worse, why not always use stochastic methods? There are two reasons. First, only the simplest stochastic equations can be solved explicitly. If, in population dynamics or in epidemiology, we construct stochastic equations which are remotely realistic, they will certainly be insoluble. The difficulties are such that stochastic theory is one of the few areas of biomathematics of any intrinsic interest to professional mathematicians, as distinct from biologists. Such difficulties may now be circumvented, not very satisfactorily, by computer simulation. The second reason is the sheer difficulty of identifying what the stochastic equations should be. Since population biology is so complex, it is hard enough to do the field work needed to specify deterministic equations which describe a given population process with any realism; the extra work needed to describe the possible variations is usually prohibitive. As discussed above, it is often unnecessary too.

Note

1. Lotka himself *did* test his theories against empirical data: but the only data available in his time came, not from wild populations, but from laboratory experiments, e.g. on flour beetles kept at constant temperature in bottles containing homogenized flour. Populations grown in artificially simple conditions often do conform to simple mathematical equations, as in Example 13.2.

12   A warning

This book treats statistical analyses as tools. If the biological answer is already obvious, there is no need to do the analysis. But that utilitarian point of view does not condone sloppiness. The tools must be used carefully. It is no use doing an analysis if you don't do it correctly. Before the advent of computers, analyses were done twice over—preferably by different people. Even then, the two people occasionally made the same arithmetic mistake. All calculations made with hand calculators need to be checked. People who do not normally make checks are surprised to find how often they make mistakes. It is no use assuming that mistakes will not occur; you should accept that they will, and take steps to detect them. Fortunately, computers make many fewer mistakes than do humans. The weak point is the input of data. If the data fed into the computer are wrong, the answer will be wrong too. The best check is to make the computer print out a copy of the data, and then to compare those data with the original records.

Using computers, we do not get the same intimate acquaintance with the data that we get when using desk calculators. Not only must the data be carefully checked, but the data checks build into good computer programs should be examined to detect aberrant values. Some people argue that since the original data are subject to errors of measurement, there is no need to worry if further errors are introduced during the analysis. That careless attitude leads to trouble. The extra errors will inflate the residual mean square, so reducing the apparent accuracy of the results. And once data checks have been compromised, there can be no assurance that the introduced errors are unimportant—they may seriously affect estimates of means or regressions. Every analysis must be done rigorously. For example, if you decide to predict y from x by a straight line, you should use regression analysis. Lines drawn by eye can differ surprisingly from the true regression line; after all, there are *two* distinct regression lines, namely y on x and x on y (Chapter 3). Inefficient methods of analysis can give answers which the data, treated properly, actually contradict. So if an analysis is superfluous, don't do it; but if it is necessary, do it well.

13 Curve-fitting

So far we have been dealing mainly with linear models (Chapter 4). But in biology, straight lines are rather rare. Very often a simple transformation will convert a curve to something that for practical purposes is, near enough, a straight line (Example 3.1). There is a good reason for using linear methods wherever possible. In many areas of applied mathematics—control theory, econometrics and thermodynamics—the theory is very strong for linear applications and painfully weak for non-linear. Statistics is no exception. It is certainly possible to study particular types of curve, e.g. exponentials or polynomials, in some detail; but the class of 'all possible non-linear curves' is too broad to permit anything but vague theoretical generalizations.

Yet sometimes we do need to fit a curve to a set of data, especially where no obvious transformation to linearity exists. Such problems do not occur very often in everyday statistical analysis, but nowadays they arise frequently when we make computer simulation models. We shall not discuss here the construction and use of simulation models, but this chapter will consider curve-fitting in general. The problem is this: given a set of data, we wish to fit a curve to predict appropriate values of y from the corresponding values of x. Usually there is no theoretical reason to prefer any particular type of curve. So the problem has two parts: first to choose a suitable type of curve, and second to estimate the appropriate parameters.

13.1. Choosing the type of curve

In the past, many statisticians have used 'smoothing' techniques to remove accidental irregularities from the data. A modern version of such techniques is 'splines', which consist of short lengths of polynomial curves fitted to local parts of the data range. Such techniques are little used in biology, where there is no reliable way of distinguishing genuine biological effects from accidental irregularities. Here we are not merely smoothing, but choosing an appropriate type of mathematical function and fitting its parameters to the data.

There are no rules for choosing curves. Choice of an appropriate formula is largely a matter of experience. Where the data show a steady, if non-linear, progression as in Fig. 13.1(a), it is usually quite easy to find a suitable form of curve to fit them. Indeed it is often quite easy to find

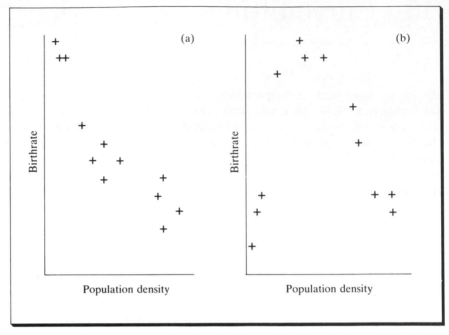

Fig. 13.1. Curve-fitting: (a) a steady progression; (b) a more tortuous shape.

several different curves, and different mathematicians tend to favour different types of curve. Provided there is no theoretical reason to prefer one over another, and provided we do not want to extrapolate along the curve beyond the range of observations, any one of those curves may be adopted. For example, in Fig. 13.1(a) we might try an exponential $y = a + \exp[c(b - x)]$, a parabola $y = a + c(x - b)^2$ or a hyperbola $y = a + c/(x - b)$. All those formulae have three parameters a, b and c. Their actual values would be very different in each case, but they would still be doing the same jobs: a and b serving to locate the scales of y and x, while c determines the amount of curvature. The three formulae might all fit equally well, but they would diverge sharply outside the range of observations shown in Fig. 13.1(a). It would be quite easy to find a transformation, perhaps the mathematical equivalent of one of these same formulae, to convert Fig. 13.1(a) into a straight line.

But sometimes the data follow a more tortuous path, as in Fig. 13.1(b). We can then try to find a composite formula whose different parts will take care of different parts of the curve. At the right-hand side of Fig. 13.1(b) the values of y seem to be declining gradually towards zero, which suggests an exponential decline, $y = e^{-bx}$. The initial rise of y might then be represented by a power of x, $y = x^a$, or perhaps by a positive exponential, $y = e^{ax} - 1$ (the value 1 is subtracted because y has to be 0 when $x = 0$). Putting the parts together, we might try $y = cx^a e^{-bx}$ or $y = c(e^{-ax} - e^{-bx})$; the first formula is rather easier to fit (Example 13.1). There is of course no guarantee that either formula shall fit a given set of data satisfactorily.

A useful key to different types of equation, originally devised by K. E. F. Watt, appears in Southwood (1966). Although Watt's curves are quoted as differential equations, they can all be solved to give an explicit relation between y and x. If no suitable formula for a curve can be found, it is always possible to fit a polynomial (Chapter 4)—and indeed a quadratic or cubic polynomial can be perfectly satisfactory. But a polynomial fitted to a rather tortuous curve like Fig. 13.1(b) must include many more coefficients than the three (a, b, c) of $y = cx^a e^{-bx}$; so analytic functions of the latter type, where applicable, are preferable, partly for their mathematical elegance but mainly because of the dangers of extrapolation mentioned in the next paragraph. Finally it is always possible to avoid the problem of curve-fitting by tabulating a series of values of y and interpolating between them. There are various methods of interpolation, but each is equivalent to fitting a particular type—linear, quadratic etc.—of curve to a small local region of the whole curve. Interpolation is evidently not very satisfactory when the data points jump about, as in Fig. 13.1, and the previously mentioned 'splines' were invented to smooth out local irregularities.

Chapter 4 mentioned the dangers of extrapolating outside the range of the data. But we may be fitting a curve because we want to use it in a simulation model, and once that simulation model is made we may ask 'What would happen if ... ?' In other words, we shall present new and hypothetical situations to the computer. Very often those hypothetical situations are simply new assortments within the observed range of parameters, but sometimes they involve some extrapolation beyond that range. Extrapolations a short distance beyond the range may reasonably be accepted, but the grosser the extrapolation, the more sceptical we must be. The dangers of extrapolation are especially great in a high-order polynomial, which sometimes bends about disconcertingly, just outside the range of the data to which it has been fitted. That is why polynomials of order higher than quadratic or cubic are best avoided, except as a last resort.

13.2. Determining parameters

Once we have chosen a suitable type of curve, we have to find appropriate values for its parameters to suit the data. Sometimes it is possible to transform to linearity, and use linear regression (Example 13.1). Sometimes no such transformation exists, in which case we can use non-linear regression (Chapter 4) or its equivalent, trial-and-error linear regression. Computer programs for non-linear regression work on the same trial-and-error principle used in Example 13.2, but they work automatically so that the user has no control over the operation. As Example 13.2 shows, a method which pays too much attention to one particular part of the curve may leave another part unconsidered. Once a curve has been fitted, by whatever method, it should always be plotted against the original data, partly to see if the mathematical formula is appropriate in the first place, and partly to make sure that the fit is everywhere good. In some cases an automatic non-linear regression program will not achieve convergence, i.e. it will not produce one definite answer. That may mean that it is being asked to fit a wildly inappropriate curve to the data, but it may merely mean that the particular program cannot cope adequately with the particular set of data. Such ambiguities do not arise when we ourselves control the curve-fitting operation, as in Example 13.2.

It is tempting, once a formula has been found, to interpret its terms biologically: 'this term represents such-and-such a biological process, while that one ...' It is very dangerous to do so, unless the particular type of formula was chosen *a priori* for theoretical reasons—and then only if the theory is apt. If a set of data conforms to some curve, the underlying biology cannot be deduced with any certainty, any more than it can be deduced from an observed distribution (Chapter 9). On the other hand, if we observe that (say) $y = ae^{kt}$, y must obey the equation $dy/dt = ky$, even though the reason may be obscure.

Examples 13

13.1. To fit a curve to the net number of eggs produced per day by a population of insects, as a function of population density.

Number of eggs (y)	4	14	23	35	24	15	9	2	2
Density (x)	20	48	87	210	280	410	500	600	700

The plot of y against x looks rather like Fig. 13.1(b).
(a) The curve is very skew. The logarithmic transformation expands the lower part of the scale of measurement (Chapter 5), and so the plot of y against $\log(x)$ might be shaped like a Normal probability curve, in which case y will be proportional to $\exp[a(\log(x) - m)^2]$ for some values of a and m. This model is converted to the standard multiple regression eqn (3.5) by writing

$$\log(y) = a[\log(x)]^2 + b\log(x) + c.$$

Find a, b and c by multiple regression of $\log(y)$ on $\log(x)$ and $[\log(x)]^2$, using natural logarithms for comparison with (b).
(b) As stated in the text, $y = cx^a e^{-bx}$ might be suitable. When x is near zero, e^{-bx}

is nearly 1 and the curve is dominated by x^a. When x is large, the exponential outweighs x^a in importance. If $y = cx^a e^{-bx}$, the relation is made linear by

$$\log(y) = \log(c) + a \log(x) - bx \text{ (using natural logarithms)}.$$

Find a, b and c by multiple regression of $\log(y)$ on $\log(x)$ and x. Try plotting $y = cx^a$, $y = e^{-bx}$, and $y = cx^a e^{-bx}$. Is (a) or (b) preferable in this case?

13.2. An exercise in curve-fitting by non-linear regression. The data of Example 3.1 show exponential growth which seems to slow down towards the end. The equation for pure exponential growth is $dN/dt = rN$ (Watt's curve A24 in Southwood 1966). We may therefore try $dN/dt = rN(A - N)$, where A is the asymptote; as N approaches A, the rate of growth declines. This is Watt's equation A22. Its solution takes the form $N = 1/(a + be^{-kt})$. This is a sigmoid curve which starts at $N = 1/(a + b)$ when $t = 0$ and climbs towards the asymptote $N = 1/a$ as t tends to infinity. In this example the values of N are generally large, and it is convenient for reasons of scale to use the equivalent $N = 100/(a + be^{-kt})$. Methods (a) and (b) will compare two different ways of fitting this curve to the data.

(a) If $N = 100/(a + be^{-kt})$ it follows that $100/N = a + be^{-kt}$, so that the regression of $100/N$ on e^{-kt} should be linear for some value of k. Form a new variate $y = 100/N$ and new variates $x = e^{-kt}$ for trial values of $k = 0.0115$, 0.012, 0.0125, 0.013 and 0.0135. Regress y on each of the xs and show that the correlation between y and x is greatest when $k = 0.0125$. From the corresponding regression equation deduce a formula for N in terms of t. Plot the data and the fitted curve. This formula gives a very bad representation of the growth curve, because the transformation $x = e^{-kt}$ makes the small values of t overwhelmingly important, at the expense of later values.

(b) If $N = 100/(a + be^{-kt})$ it follows that $\log(100/N - a)$ should be a linear function of t for some value of a. If the growth curve is purely exponential, $a = 0$. Form new variates $y = \log(100/N - a)$ for trial values of $a = 0.003$, 0.004, 0.005, 0.006 and 0.007. Regress each y on t and show that the correlation between y and t is greatest when $a = 0.005$. From the corresponding regression equation, deduce a formula for N in terms of t. Plot the data and the fitted curve. This formula gives an acceptable representation of the growth curve, because the regression of y on t pays equal attention to all values of t. But the formula takes no fewer than three parameters to represent only ten original data points, and it is still doubtful whether a differs from zero, i.e. whether the growth curve is not purely exponential.

13.3. Some standard types of curve
(a) The exponential, $y = e^{kx}$ (Watt's A24 and A30 in Southwood 1966). Plot the four curves corresponding to $k = 2$, 1, -1 and -2 for values of $x = 0$, 1, 2, 3, 5, 10. What happens if $k = 0$?
(b) The power, $y = x^a$ (Watt's A20). Plot the four curves corresponding to $a = \frac{1}{2}$, 1, $\frac{3}{2}$ and 2 for values of $x = 0$, 1, 2, 3, 5, 10. If $a \le 1$ the curve rises directly from the origin; if $a > 1$ it touches the x-axis at the origin. What happens when $x < 0$ if a is (i) even, (ii) odd and (iii) a fraction?
(c) The sigmoid—Example 13.2.
(d) The logarithm (Watt's A28 and A31)—Example 5.1.

14   Outline of Methods

This chapter sketches the process used to tackle any particular problem. Bailey's (1981) summary tells how to choose an appropriate method of analysis. Bailey's is perhaps the best of the very large number of statistical cookery books on the market, because it tries to tell you not only when to use a given method, but when not to. The following outline attempts the impossible. Biological research cannot be done by rote, and analysis of data cannot be reduced to a mechanical set of rules. Nor can all the profound wisdom packed into this book be reduced to a few recipes!

1. First specify, as explicitly as possible, the biological questions to be asked (Chapter 8). This does not mean that you can't consider other questions which may suggest themselves later (Chapter 2).

2. If relevant data do not already exist, choose the appropriate methods of experimentation or observation. The criteria are relevance (the biological effect under examination must not be confounded with other, unwanted effects—Chapter 8), efficiency (the greatest accuracy at least cost) and practical convenience. Consider how big a sample is needed to give the desired accuracy (Chapter 8). Sometimes it is impossible or impracticable to answer the chosen biological questions, in which case return to point (1). The design of a large experiment is a job for a statistician. It is up to the biologist to make sure that the statistician understands the problem. Only too often, the statistician takes for granted things which are biologically absurd.

3. The method of analysis of the data will depend on the choice of an underlying statistical model (Chapter 4). We usually use standard methods of analysis which depend on standard models of additive effects and linear regressions (Chapters 2–4). It is often possible to transform data to satisfy those models, at least approximately (Chapter 5). The model used to analyse a complicated experiment will correspond to the design of that experiment. Consider how you will analyse the data *before* you take the observations.

4. The analysis will usually involve estimation of some parameters, e.g. means or regression coefficients, by least squares (Chapter 1) or maximum likelihood (Chapter 7).

5. Use the estimated values to see if the model agrees well enough with the data. This may be done graphically (Chapter 3) or by a special

goodness-of-fit test. Often there is no need to examine the agreement very closely, because we rely on the robustness (Chapter 4) of ordinary statistical methods.

6. Consider whether the biological story which the statistics tell makes sense. If not, either you have done something wrong or you are on the trail of a momentous discovery. Consider the size of the biological effects observed, to see if they are worth bothering about. If the story is worth pursuing, return to point (1).

If you are in doubt about the suitability of a proposed method of analysis to answer some chosen question, simply translate that analysis into its underlying model (Chapter 4) and distribution of remainders. It will then be easy to decide whether the analysis is suitable. The commonest type of analysis examines the means of one or more samples, usually via an analysis of variance. The variance-ratio (F-) test can be used to compare several means; that same test, or the equivalent t-test, can compare two means, or one mean and its theoretical value (Chapter 2). If the data are whole-number counts, they may need transformation (Chapters 5 and 11). A transformation may also be advisable for data which can vary continuously, e.g. direct measurements or percentages, but quite often the standard assumptions of additive treatment effects and approximately Normal residuals are adequate. Two-way tables should be made orthogonal wherever possible, because non-orthogonal tables are difficult to interpret (Chapter 2).

The next most common type of analysis examines the relation between two or more variates. Correlation measures the strength of the association, i.e. the degree to which one variate may be predicted from another, while regression gives the actual prediction formula (Chapter 3). Both correlation and regression assume linearity. Multiple regressions may be difficult to interpret, and in any case, correlations and regressions never prove causality (Chapter 3). Only deliberately designed experiments can prove causality (Chapter 8). The analysis of experimental results is dictated by the design of the experiment, i.e. the analysis examines those questions which are built into the experiment.

Simple analyses of *counts* are usually based on the Binomial or Poisson distributions: Binomial if the count has a top limit, e.g. 'of 100 humans, 47 are males' and Poisson if there is no limit. The two distributions are related (Note 2, Chapter 6). Values of χ^2 calculated from contingency tables and other sets of counts are really weighted sums of squares (Chapter 1). Before comparing a calculated value of χ^2 with its theoretical (tabulated) value, make sure that the underlying assumptions are justified (Chapter 6). More complicated analyses of counts, involving two or more cross-classifications, may involve transformation and weighting (Chapter 5) or a 'generalized linear' analysis (Chapter 7).

Examples 14

The following examples are not restricted to the content of this chapter, but invoke arguments drawn from all parts of the book.

14.1. A female *Aphidius smithi* is given 180 aphids to parasitize. She lays eggs one at a time. Each aphid is dissected and the number of parasite eggs counted. Are some aphids more attractive than others?

Eggs per aphid	0	1	2	3	4	5	6
Number of aphids	15	94	48	15	6	1	1

14.2. A zoologist measures the lengths of the animals in two representative samples of *Lumbricus terrestris*. One sample is from S.E. England and one from W. Scotland. Is there a difference in length for the species between the two regions? (data from Sprent 1970).

Lengths (cm)

S.E. England	13.2	12.1	12.5	13.4	13.7	12.9	13.2	12.8	13.4
W. Scotland	11.3	12.1	11.4	12.2	11.3	12.5	11.2	11.3	

14.3. A sample of 3459 people gave the following blood group frequencies:

A	B	AB	O
1546	297	113	1503

A and B are both dominant to O. Are the data consistent with random association of alleles?

14.4. We have a sample containing N values of the variables x and y. The values are all positive: when plotted, they lie roughly on a straight line through the origin. Under what circumstances would you estimate the slope of the line by

(a) $\sum xy / \sum x^2$ (b) $\sum y / \sum x$ (c) $\sum (y/x)/N$

14.5. An algal filament consists of a chain of single giant cells joined end to end. The filament grows by cell division at one end only. There is an ample stock of filaments. We want to do an experiment on the effect of a metabolic inhibitor on the rate of uptake of potassium into the cell. It is known that (a) the rate of uptake of one cell does not affect the rate of uptake of any other cell, (b) the inhibitor has no effect unless it is introduced into the cell by micropipette, which can be done without damaging the cell. The rate of uptake is measured by placing the filament for a given period of time in a bath containing radioactive potassium ions, and then using autoradiography to count the number of decays per cell. Individual cells can be identified on the autoradiograph. How would you do the experiment?

14.6. The following data show the result of an experiment on the effect of two supposedly soporific drugs A and B, in producing sleep. (a) Compare the average effects of A and B. (b) Compare the average effects of A and B, supposing that the data were obtained by using different patients for the two drugs.

Patient	Additional hours of sleep with	
	A	B
1	+ 0.7	+ 1.9
2	− 1.6	+ 0.8
3	− 0.2	+ 1.1
4	− 1.2	+ 0.1
5	− 0.1	− 0.1
6	+ 3.4	+ 4.4
7	+ 3.7	+ 5.5
8	+ 0.8	+ 1.6
9	0.0	+ 4.6
10	+ 2.0	+ 3.4

References

Bailey, N. T. J. (1967). *The mathematical approach to biology and medicine.* Wiley, New York.

Bailey, N. T. J. (1981). *Statistical methods in biology* (2nd edn). Halstead/Wiley, New York.

Caughley, G. (1977). *Analysis of vertebrate populations.* Wiley, London.

Cochran, W. G. and Cox, G. M. (1957). *Experimental designs.* Wiley, New York.

Connell, J. H. and Sousa, W. P. (1983). On the evidence needed to judge ecological stability or persistence. *American Naturalist,* **121**, 789.

Dingle, H. and Hegmann, J. P. (ed.) (1982). *Evolution and genetics of life histories.* Springer, New York.

Ehrlich, P. R. and Murphy, D. D. (1981). The population biology of checkerspot butterflies. *Biologisches Zentralblatt,* **100**, 613.

Falconer, D. S. (1981). *Introduction to quantitative genetics* (2nd edn). Longman, London.

Fisher, R. A. (1918). The correlations between relatives on the assumption of Mendelian inheritance. *Transactions of the Royal Society of Edinburgh,* **52**, 399.

Fisher, R. A. (1930). *The genetical theory of natural selection.* Clarendon, Oxford.

Fisher, R. A. (1956). *Statistical methods and scientific inference.* Oliver and Boyd, Edinburgh.

Fisher, R. A. (1971). *The design of experiments* (9th edn). Hafner/MacMillan, New York.

Fisher, R. A. and Yates, F. (1963). *Statistical tables for biological, agricultural and medical research* (6th edn). Oliver and Boyd, Edinburgh.

Gilbert, N. (1982). Comparative dynamics of a single-host aphid. *Journal of Animal Ecology,* **51**, 469.

Gilbert, N. (1988). Control of fecundity in *Pieris rapae. Journal of Animal Ecology,* **57**, 395.

Goel, N. S., Maitra, S. C. and Montroll, E. W. (1971). On the Volterra and other nonlinear models of interacting populations. *Review of Modern Physics,* **43**, 231.

Greig-Smith, P. (1983). *Quantitative plant ecology* (3rd edn). Blackwell, Oxford.

Harper, J. L. (1977). *Population biology of plants.* Academic, London.

Healy, M. J. R. (1988). *Glim: an introduction.* Clarendon, Oxford.

Jeffreys, H. (1939). *Theory of probability.* Clarendon, Oxford.

Lwin, T. and Maritz, J. S. (1982). An analysis of the linear-calibration controversy from the perspective of compound estimation. *Technometrics* **24**, 235.

Mather, K. and Jinks, J. L. (1982). *Biometrical genetics* (3rd edn). Chapman and Hall, London.

McCullagh, P. and Nelder, J. A. (1983). *Generalized linear models.* Chapman and Hall, London.

McIntosh, R. P. (1985). *The Background of Ecology.* Cambridge University Press.

Pearce, S. C. (1983). *The agricultural field experiment.* Wiley, Chichester.

Ricker, W. E. (1973). Linear regressions in fishery research. *Journal of the Fisheries Research Board of Canada*, **30**, 409.

Southwood, T. R. E. (1966, 1978). *Ecological methods* (1st and 2nd edns). Methuen/Chapman and Hall, London.

Sprent, P. (1970). Some problems of statistical consultancy. *Journal of the Royal Statistical Society*, **A133**, 139.

Tukey, J. W. (1949). One degree of freedom for non-additivity. *Biometrics*, **5**, 232.

Tukey, J. W. (1954). Causation, regression and path analysis. In *Statistics and mathematics in biology* (ed. O. Kempthorne). Iowa State College Press.

Wang, Y. H. and Gutierrez, A. P. (1980). An assessment of the use of stability analyses in population ecology. *Journal of Animal Ecology*, **49**, 435.

Wilkinson, G. N., Eckert, S. R., Hancock, T. W. and Mayo, O. (1983). Nearest neighbour (NN) analysis of field experiments. *Journal of the Royal Statistical Society*, **B45**, 151.

Yates, F. (1937). *The design and analysis of factorial experiments*. Technical communication 35, Commonwealth Bureau of Soil Science, Harpenden.

Yates, F. (1947). The analysis of data from all possible reciprocal crosses between a set of parental lines. *Heredity*, **1**, 287.

Yates, F. (1955). A note on the application of the combination of probabilities test to a set of 2×2 tables. *Biometrika*, **42**, 404.

Answers to examples

Example 1.1

The point of this example is that some people think algebraically, i.e. in symbols, and others geometrically, in pictures. If you think in pictures, the graph will show how the analysis of variance works.

Example 1.2

(a) *By calculus*: differentiating $\sum(y-m)^2$ with regard to m and equating the differential to zero, $\sum(y-m)=0$, i.e. $m=\bar{y}$.

(b) *By algebra*: let $m=\bar{y}+a$ say. Then

$$\sum(y-m)^2 = \sum(y-\bar{y}-a)^2$$
$$= \sum[(y-\bar{y})^2 - 2a(y-\bar{y}) + a^2]$$
$$= \sum(y-\bar{y})^2 + Na^2 \text{ since } \sum(y-\bar{y})=0.$$

Therefore $\sum(y-m)^2$ is least when $a=0$, i.e. when $m=\bar{y}$.

Example 1.3

$$\sum(y-\bar{y})^2 = \sum(y^2 - 2y\bar{y} + \bar{y}^2)$$
$$= \sum y^2 - 2\bar{y}\sum y + N\bar{y}^2$$
$$= \sum y^2 - 2\bar{y}(N\bar{y}) + N\bar{y}^2$$
$$= \sum y^2 - N\bar{y}^2.$$

Example 2.1

Analysis of variance:

	Degrees of freedom	Mean square
Between sexes	1	172.8
Residual	6	1.2

It would be possible instead to treat males and females quite separately:

	Degrees of freedom	Sum of squares	Mean square
Variance of males	4	5.2	1.3
Variance of females	2	2.0	1.0

Unless there is reason to suppose that males are intrinsically more variable than females, or vice versa, the two estimates of variance are combined to obtain greater accuracy (more degrees of freedom). Then the combined sum of squares $5.2 + 2.0$ has $4 + 2$ degrees of freedom, giving the mean square 1.2 (with 6 degrees of freedom) which appears in the analysis of variance. This mean square is used to calculate the following variances on the assumption that in future we shall always distinguish males from females, and allow for the difference between them. By taking the trouble to distinguish the two sexes, we eliminate that part of the overall variability which arises from the difference between male and female.

Mean of males	38.6 with variance 0.24	
Mean of females	29.0	0.40
Overall mean	35.0	0.15
Average of male and female means	33.8	0.16
Difference between male and female means	9.6	0.64

The average of male and female means has greater variance, i.e. is less accurate than, the overall mean. That must always be so, unless the sample numbers of males and females are the same. But there is evidently a real difference between males and females—the estimate of the difference, 9.6, is twelve times its standard error $\sqrt{0.64}$. Therefore the overall mean composed of 5 males : 3 females is rather meaningless. It estimates the mean of all possible samples which happen to contain a 5 : 3 sex ratio. On the other hand, the average of male and female means estimates the mean of all samples, or of a whole population, with a 1 : 1 sex ratio.

Example 2.2

The value 4.37 is obviously wrong. No pig can grow 4 cm per day, even for one day. If we use nonsensical data, the answer will be nonsense.
(a) Omitting the value 4.37 altogether, the analysis of variance is:

	Degrees of freedom	Mean square
Between lots	2	0.10326
Residual	10	0.00410

$$\bar{y}_1 = 1.08 \text{ with standard error } 0.032$$
$$\bar{y}_2 = 1.28 \qquad\qquad\qquad 0.032$$
$$\bar{y}_3 = 1.38 \qquad\qquad\qquad 0.029$$
$$\bar{y}_2 - \bar{y}_1 = 0.20 \qquad\qquad\qquad 0.045$$
$$\bar{y}_3 - \bar{y}_2 = 0.10 \qquad\qquad\qquad 0.043$$
$$\bar{y}_3 - (\bar{y}_1 + \bar{y}_2)/2 = 0.20 \qquad\qquad\qquad 0.036$$

(b) Assuming that 4.37 is a misrecording for 1.37 (a reasonable guess in view of the slight amount of variation) and substituting that value, the analysis of variance becomes:

	Degrees of freedom	Mean square
Between lots	2	0.10889
Remainder	11	0.00374

$$\bar{y}_1 = 1.08 \text{ with standard error } 0.031$$
$$\bar{y}_2 = 1.28 \qquad\qquad\qquad 0.031$$
$$\bar{y}_3 = 1.38 \qquad\qquad\qquad 0.025$$
$$\bar{y}_2 - \bar{y}_1 = 0.20 \qquad\qquad\qquad 0.043$$
$$\bar{y}_3 - \bar{y}_2 = 0.10 \qquad\qquad\qquad 0.039$$
$$\bar{y}_3 - (\bar{y}_1 + \bar{y}_2)/2 = 0.20 \qquad\qquad\qquad 0.033$$

The genetical conclusions are the same in either case. The comparison $\bar{y}_2 - \bar{y}_1$ shows that Landrace grow rather faster than Large White, and the other two comparisons show that hybrids grow faster than the parental average, but not significantly faster than the best parent.

(c) If the deviant value 4.37 is retained, $\bar{y}_3$ becomes 1.88 and the residual mean square increases to 0.6801. The mean changes by 36 per cent and the mean square by 16 488 per cent. This shows that the variance is very much less robust than the mean (Chapter 4).

Example 2.3

Analysis of variance:

	Degrees of freedom	Mean square
Between treatments	1	4.8000
Between nests within treatments	8	0.4743
Between chicks within nests	20	0.1280

Why are there not 9 degrees of freedom between the ten nests?—because one of those degrees of freedom has already been accounted for as the difference between treatments. After *two* treatment means have been fitted to *ten* nests, there are only 8 degrees of freedom left between nests—4 between the five 'control' nests and 4 between the five 'extra food' nests. Similarly the 20 degrees of freedom 'between chicks within nests' are made up of 2 degrees of freedom within each of 10 nests.

On average, the chicks given extra food gained more weight (how much?). The treatment 'extra food' is given to whole nests, not to individual chicks within a nest. The experimenter has no control over which chicks get how much food. So the nest is the experimental unit, and the comparison between treatments is a comparison between two sets of *nests*. If extra food had no effect, we should expect the 'between treatments' mean square to be the same as the 'between nests' mean square. Therefore, to test the significance of the treatment difference, we must use the 'between nests' mean square as residual, not the 'between chicks within nests' mean square (Chapter 6). If the 'between nests' mean square had turned out to be the same size as the 'between chicks within nests' mean square, we could assume that there were no consistent differences between nests (within any one treatment), so that 'between nests' and 'between chicks' were estimating the same residual variance. It would then be valid to calculate a combined residual with 28 degrees of freedom. In this case, 0.4743 is considerably—even if not significantly—bigger than 0.1280, and so the two mean squares should not be combined. Therefore, to compare the two treatments we might just as well have recorded nest totals rather than individual chicks, because the nest is the relevant experimental unit.

Example 2.4

Analysis of variance:

	Degrees of freedom	Mean square
Between sexes	1	24.50
Between species	2	328.22
Interaction	2	48.67
Remainder	12	5.67

The variance ratio 24.50/5.67 shows that males are on average faster than females, but the variance ratio 48.67/5.67 shows that the difference varies from species to species. The original data show that male kangaroos are considerably faster than females, but the two sexes of cheetah and greyhound are about equally fast. This makes biological sense, because male kangaroos are larger than females, who are almost always carrying young (although a female kangaroo in real danger will jettison her joey). Cheetahs are faster than kangaroos, which are faster than greyhounds (by how much?).

Example 2.5

(a) Analysis of variance:

	Degrees of freedom	Mean square
Between nationalities	2	19.29
Residual	15	5.09

Australians apparently drink more than others.

(b) Analysis of variance:

	Degrees of freedom	Mean square
Nationalities	2	19.29
Sexes (adj. nationalities)	1	58.43
Nationalities (adj. sexes)	2	0.31
Sexes	1	96.39
Interactions	1	0.14
Remainder	13	1.37

The nationalities mean square is necessarily the same as in (a).
Nationalities (adj. sexes) mean square does not signify, and so the
apparent differences between nationalities may be ascribed to the
fact that the Australians were all male. Inspection of the data shows
that the interpretation is reasonable. Why is there only one degree of
freedom for interactions?

Example 2.6

If a mean square is large, its sum of squares must correspondingly be
large. Since 'columns' and 'rows adj. columns' sums of squares are both
large, their sum, i.e. the sum of squares for rows and columns together,
must be large too. Therefore, since 'rows' sum of squares is not large,
'columns adj. rows' must be large. Since both 'columns' and 'columns adj.
rows' mean squares are large, there are real differences between
columns. 'Rows adj. columns' mean square is large, meaning that there
are some real differences between rows which cannot be explained away
as indirect effects of columns. By chance, those row effects cancel out
when mixed with the indirect columns effects in the 'rows' mean square.
Such a situation cannot arise in an orthogonal table, where 'rows' and
'rows adj. columns' mean squares are identical.

Example 2.7

The residual sum of squares, minimized by fitting both row and column constants, cannot exceed the residual sum of squares minimized by fitting row constants only. Therefore the sum of squares *accounted for* cannot be less.

Example 2.8

By definition V = average of $(y-m)^2$
$$= \text{av}(y^2 - 2my + m^2) = \text{av}(y^2) - m^2.$$

Therefore $\text{av}(y^2) = m^2 + V$. Similarly, $\text{av}(\bar{y}^2) = m^2 + V/N$ since the mean and variance of $\bar{y}$ are m and V/N. Therefore

$$\text{av}[\sum(y-\bar{y})^2] = \text{av}(\sum y^2 - N\bar{y}^2)$$
$$= N\text{av}(y^2) - N\text{av}(\bar{y}^2)$$
$$= N(m^2 + V) - N(m^2 + V/N)$$
$$= (N-1)V.$$

Example 2.9

The weighted mean $\bar{y} = \sum a_i y_i$ where $a_i = w_i/\sum w_i$. Its variance $= \sum a_i^2 V_i$ which reduces to $1/\sum w_i$ since $V_i = 1/w_i$.

Example 3.1

(a) Count $= -2426 + 29.2$ (time), a significant regression but a poor fit.
(b) $\log_e$ (count) $= 3.511 + 0.0152$ (time), a much better representation but perhaps still not perfect. The growth of bacteria was exponential for 300 minutes, but thereafter seems to slow down (Example 13.2).

Example 3.2

The regressions are the same for adult males and females, and so those two categories may be combined. Then the regression of weight on distance for adults is

$$\text{weight} = 4.06 - 0.0026 \text{ (distance)},$$

giving predictions of 2.8 g and -1.1 g at 500 miles and 2000 miles. The negative figure is an extrapolation outside the range of the data, and shows that the bird could not fly 2000 miles non-stop. The regression of distance on weight for juveniles is

$$\text{distance} = 1194 - 232 \text{ (weight)},$$

giving a prediction of 730 miles flown by a bird weighing 2.0 g.

Example 3.3

Consider the last part first, because it's simpler. If V is the variance of individual values of y, the mean of y in block 1 is $\bar{y}_1 = \sum_1 y/N_1$ with variance V/N_1 and therefore with weight $w_1 = N_1/V$. Similarly for the second block. The weighted mean is $(w_1 \bar{y}_1 + w_2 \bar{y}_2)/(w_1 + w_2)$ which, on substituting the values of w and $\bar{y}$, becomes $(\sum_1 y + \sum_2 y)/(N_1 + N_2)$. So the weighted mean, which is the most accurate combination of the two block means, equals the overall mean obtained by lumping the two blocks together as one.

The situation is slightly more complicated for regressions, which involve two parameters a and b (eqn (3.4)). Here

$$b_1 = \sum_1 (y - \bar{y}_1)(x - \bar{x}_1)/\sum_1 (x - \bar{x}_1)^2$$

with variance $V/\sum_1 (x - \bar{x}_1)^2$, and similarly for b_2. The weighted mean value of b is

$$[\sum_1 (y - \bar{y}_1)(x - \bar{x}_1) + \sum_2 (y - \bar{y}_2)(x - \bar{x}_2)]/[\sum_1 (x - \bar{x}_1)^2 + \sum_2 (x - \bar{x}_2)^2].$$

This is not the same as the regression obtained by lumping the two blocks together, because the weighted mean value of b is calculated using the separate block means $\bar{x}_1$, $\bar{x}_2$, $\bar{y}_1$, $\bar{y}_2$, not the overall means. In other words, the regression slopes from the two blocks are combined but the block means, and therefore the intercepts, are kept separate.

Example 3.4

$x_1 = 1$ for Large Whites and 0 otherwise; $x_2 = 1$ for Landrace and 0 otherwise. Then the regression of growth rate on the dummy variates is $y = 1.382 - 0.302x_1 - 0.097x_2$, giving $y = 1.080$ for Large White ($x_1 = 1$, $x_2 = 0$); $y = 1.285$ for Landrace ($x_1 = 0$, $x_2 = 1$); and $y = 1.382$ for the cross ($x_1 = x_2 = 0$). The analysis of variance for the regression is identical to that for the one-way analysis.

Regression on the single dummy variate would absorb only one degree of freedom, and would require that $y = a + 2b$ for the cross, $y = a + b$ for

Landrace and $y = a$ for Large White, i.e. it would arbitrarily require that the difference between Landrace and Large White must equal that between the cross and Landrace.

Example 3.5

The single regression of 'number of babies' on calendar year leaves a residual mean square 28 595 (11 degrees of freedom). The multiple regression on years and nests leaves a residual mean square 28 422 (10 degrees of freedom). The reduction is negligible, and certainly not significant (variance ratio = 30 326/28 422 with 1 and 10 degrees of freedom). The apparent correlation between nests and babies is an indirect effect of calendar years.

Example 3.6

(a) Correlations between fleas and fish = 0.686 with 6 degrees of freedom. And the data prove neither assertion. (Those two statements are not connected.)
(b) Correlation between cat's weight and fish = 0.288 with 6 degrees of freedom, between cat's weight and fleas = -0.394 with 6 degrees of freedom. Although these correlations are in the directions that would be expected biologically, they are not large enough to give adequate prediction. Alternatively, and equivalently, the single regressions of cat's weight on fish and on fleas do not signify.
(c) The multiple regression of cat's weight on fish and fleas gives a variance ratio (regression mean square/residual mean square) of 7.26 with 2 and 5 degrees of freedom.

Example 3.7

(c) The regression equation
$$y = a + b_1(x_1 + x_3) + b_2(x_1 - x_3)$$
is identical to
$$y = a + (b_1 + b_2)x_1 + (b_1 - b_2)x_3.$$
Therefore a multiple regression on $(x_1 + x_3)$ and $(x_1 - x_3)$ must always give the same predictions as—and its coefficients may be deduced from—the multiple regression on x_1 and x_3. Similarly (d) $y = a + b(x_1 + x_2 + x_3)$ is the same as $y = a + 3b(x_1 + x_2 + x_3)/3$, i.e. the regressions on

$x_1 + x_2 + x_3$ and on $(x_1 + x_2 + x_3)/3$ are equivalent, with one coefficient three times the other.

Example 4.1

In the extreme case, suppose that x_1 always equals x_2 and that the regression of y on x_1 is $y = a + bx_1$. Then the multiple regression of y on x_1 and x_2, i.e. on the same x-variate twice, must give precisely the same predictions, and may be written $y = a + gx_1 + (b - g)x_2$, where g can take any value we please. For whatever g may be, this equation reduces to $y = a + bx$. So the multiple-regression coefficients g and $(b - g)$ are indeterminate, but their sum is not, and the predictions are the same whatever g may be. The coefficients actually calculated by the computer will depend on how the program does the calculations. The program ought to complain that the x-variates are linearly dependent. In this numerical example, the two x-variates are very highly, but not completely, correlated. The multiple-regression coefficients are unstable and their standard errors are therefore very large, compared with the standard errors of the corresponding single-regression coefficients.

Example 5.3

All three plots are approximately linear. If any transformation is used at all, (c) is preferable because the scatter of y about the regression line remains the same over the whole range of values of x, whereas in (b) the scatter becomes greater, i.e. the residual deviations of y from the straight line increase in size as the rabbits get bigger and so the regression should, strictly speaking, be weighted to allow for the increase in residual variance.

Example 5.4

For $N = 10$ the 95 per cent confidence limit is 0.259, i.e. 25.9 per cent. For $N = 100$ the 95 per cent limit is 0.030.

Example 7.1

(a) The expected numbers are 50 males and 50 females. With one degree of freedom

$$\chi^2 = (36 - 50)^2/50 + (64 - 50)^2/50 = 7.84.$$

(b) $p = 0.64$ with 95 per cent confidence limits

$$0.64 \pm 1.96\sqrt{(0.64 \times 0.36/100)},$$

i.e. 0.546 and 0.734. This sum is best done with fractions rather than percentages, because the formula $p(1 - p)/N$ applies to fractions.

(c) The angular transform of 64 per cent is 53.1 degrees, with variance 820.7/100. Its 95 per cent confidence limits are $53.1 \pm 1.96\sqrt{8.207}$, i.e. 47.5 and 58.7 degrees, corresponding to 54.3 and 73.0 per cent. The sum is done in percentages because the angular transform is tabulated in percentages. In this example the answers to (b), (c) and (d) are all very similar.

Example 7.2

This example illustrates the tightness of the scale of percentages near 100 per cent (Chapter 5). The answers to (b), (c) and (d) now differ; (c) and (d) make better sense than (b) because the assumption in (b)—that the original data are Normally distributed—is badly wrong in this case.

Example 8.1

(a) Suppose there are N pigs per treatment. The standard error of the difference between two treatments will be $\sqrt{[0.0041(1/N + 1/N)]}$. If the difference is 0.05 cm/day and $t = 1.96$, then

$$1.96 = 0.05/\sqrt{(0.0041 \times 2/N)},$$

so that $N = 12.6$. There must be 13 pigs per treatment.

(b) Either procced as in (a), or as follows: the accuracy required is 5 times greater than in (a), therefore the sample size must be 25 times bigger, i.e. $N = 25 \times 12.6 = 315$.

Example 9.1

(a) The mean square 1.773 is estimated from a sample of 244 values of y, and so has 243 degrees of freedom. The value 1.488 is a theoretical variance, and so has infinite degrees of freedom. The variance-ratio test is approximate because it assumes the residuals $y - \bar{y}$ to be Normally distributed.

(b) The table of expected frequencies has six entries. There are two constraints imposed on those expected frequencies, namely the total must equal the sample size 244 and the mean of y must equal the observed value 1.488. Therefore χ^2 has four degrees of freedom. The χ^2 test is approximate because it assumes that the differences (observed − expected frequencies) are Normally distributed.

Example 10.1

This example reiterates the argument of Note 5, Chapter 3. The analysis of variance splits the total sum of squares into two or more parts which cannot be negative, whatever the data may be (Example 2.7). It is only when we come to interpret the results of the analysis of variance that we assume additive models and Normal distribution of remainders. So those assumptions are not part of the arithmetic method of analysis, but of the interpretation of the calculated mean squares. The genetic and environmental components of variance represent one way of interpreting the mean squares. They depend on the assumption of additivity in the model of eqn (10.1). If that model is seriously wrong, the analysis of variance itself is unaffected—in particular, the sums of squares cannot be negative—but the results cannot be interpreted in terms of an additive model, and therefore the components of variance, which are estimated from the mean squares in an attempt to interpret those mean squares, cannot be trusted and may indeed be negative.

Example 13.1

Using natural logarithms, the residual mean square of $\log(y)$ in (a) is 0.2162 and in (b) is 0.0747. So the formula (b) gives the better fit. Two formulae may be compared in this way only when they are fitted by least squares to the *same* variate, in this case $\log(y)$. We cannot directly compare the accuracies of one formula fitted to y, and another fitted to (say) $\log(y)$. But when the regression is linear and we are comparing different transformations of y on the same x-variate, we can compare the correlations instead (Example 13.2(b)).

Example 13.2

Reasonable trial values for k in part (a) and for a in part (b) are determined by inspection of the original data, or by first trying a few widely spaced values. The analysis could be prolonged by taking a further fine range of values near $k = 0.0125$ and $a = 0.005$, but the improvement would hardly be worth the trouble. In (b) the y-variate itself changes as a changes, and so we can't ask 'Which value of a gives the smallest residual mean square?', but ask instead 'Which value of a gives the greatest correlation between y and t?' The regression equation is then

$$\log_e(100/N - 0.005) = 1.2827 - 0.0173t,$$

so that

$$N = 100/(0.005 + 3.606e^{-0.0173t}).$$

The value of a, 0.005, is small compared with b, 3.606. If it were zero, we should be back at the purely exponential growth curve used in Example 3.1.

Example 14.1

The statistical method is to fit a purely random, i.e. Poisson distribution (for data in the form of counts which are not restricted by any upper limit) to 'eggs per aphid'. Here the sample size is 180 and the mean count 1.50. The goodness-of-fit may be tested by χ^2. In this case the formal test is unnecessary because the observed distribution is very far from random, containing too many 1s and too few 0s, 3s and 4s. Although the distribution is not random, it doesn't follow that some aphids are more attractive than others. If they were, we would expect to observe too few 0s and 1s, and too many 3s and 4s. Another possibility, that the parasite failed to discover some of the aphids, should produce a surplus of 0s. The most likely explanation is that the parasite discriminates against aphids which already contain a parasite egg. This makes some aphids more acceptable than others, simply because not previously parasitized. The only way to make sure is to observe directly the biological process which gives rise to the data.

Example 14.2

The statistical method is to compare the means of the two samples by a t-test or the corresponding F-test in a one-way analysis of variance. But *Lumbricus terrestris* is an earthworm: how is it possible to take a 'representative sample' of all the earthworms of this species in S.E. England? It is most unlikely that the data can be what they purport to be. This example and the previous one emphasize the need for cautious scepticism, both of the way the data were collected and of the deductions drawn from the analysis.

Example 14.3

This example illustrates the process of forming an appropriate statistical model and estimating its parameters. Let the frequencies of A, B and O alleles be p, q, r, where $p + q + r = 1$. The A group contains genotypes AA and AO, and therefore, given random association, has frequency $p^2 + 2pr$; similarly B has frequency $q^2 + 2qr$, AB $2pq$ and O r^2. As a check, these four frequencies sum to 1 because $(p + q + r)^2 = 1$. The next step is to estimate p, q and r, i.e. to find the values which maximize the log likelihood

$$1546 \log(p^2 + 2pr) + 297 \log(q^2 + 2qr) + 113 \log(2pq) + 1503 \log(r^2)$$

subject to $p + q + r = 1$. The values, which may be obtained by hand calculation or by an appropriate computer program, are $p = 0.2788$, $q = 0.0612$, $r = 0.6600$, giving expected frequencies A 1541.8, B 292.4, AB 118.0, O 1506.7, so that

$$\chi^2 = 0.0114 + 0.0724 + 0.2119 + 0.0091 = 0.305.$$

This χ^2 has one degree of freedom (starting with four frequencies A, B, AB and O, one degree of freedom is absorbed to make the expected frequencies sum to the sample total 3459, and two more to estimate p and q—the value of r follows automatically as $1 - p - q$). The expected value of χ^2 with n degrees of freedom is n, so the value 0.305 is less than its expected value 1, i.e. the data are consistent with random association of alleles.

Example 14.4

The weighted regression of y on x, passing through the origin, is $\sum w_i x_i y_i / \sum w_i x_i^2$, where $w_i = 1/V_i$ (Chapter 3). The three estimates are therefore predictive regressions of y on x with V_i (a) constant, (b) proportional to x_i, (c) proportional to x_i^2. Alternatively, (b) $\sum y / \sum x = \bar{y}/\bar{x}$, which is the slope of the line joining the origin to the mean point $(\bar{x}, \bar{y})$ of the sample, i.e. it estimates the functional relationship between x and y on the assumption that the relationship is a straight line through the origin, with both x and y subject to errors of measurement (Chapter 3).

Example 14.5

The most accurate comparisons are likely to be between cells in the same filament, rather than in different filaments. Since a filament grows at one end only, there is an age gradient of cells along the filament. Despite the statement that the inhibitor can be introduced into a cell without damaging it, there is a possibility that the process of introduction might itself affect the metabolism. Given these considerations, you might consider (a) using pairs of adjacent cells, injecting inhibitor into one chosen at random (i.e. using tables of random numbers) and a placebo into the other, or (b) using triplets of adjacent cells, injecting inhibitor into the central one and placebo into the two outside ones (or vice versa); the comparison of the central with the mean of the two outside cells eliminates the gradient, so far as it is linear. The choice between (a) and (b) depends on the amount of variation between cells and the severity of the gradient, which are probably not known before the experiment is done. The important considerations are to maximize the accuracy for a given amount of work, and to obtain a valid estimate of residual variation.

Example 14.6

(a) The design of the experiment involves giving both treatments to each patient, so that the comparison between treatments is unaffected by differences between patients. The appropriate analysis, corresponding to the design of the experiment, is a two-way analysis of variance, with patients and treatments as rows and columns. Since the variation between patients is of no direct interest, the question may be answered more simply by 'paired comparisons' i.e. a one-way analysis of the differences $(B-A)$ for each patient. This method gives the same answer as the two-way analysis. To eliminate the possibility of bias due to residual effects of the first drug persisting when the second is given, the experiment should be 'balanced', i.e. half the patients should receive A first and the other half B first.

(b) In this case the difference between the treatment means is a difference between two sets of patients, and corresponding to the different experimental design there is a different analysis—a one-way comparison of the two treatments. The answer is less accurate than in (a) but in this case we don't have to worry about possible residual effects.

Index